VOLUME ONE HUNDRED AND TWENTY ONE

INTERNATIONAL REVIEW OF NEUROBIOLOGY

Omic Studies of Neurodegenerative Disease: Part A

INTERNATIONAL REVIEW OF NEUROBIOLOGY

VOLUME 121

VOLUME ONE HUNDRED AND TWENTY ONE

INTERNATIONAL REVIEW OF NEUROBIOLOGY

Omic Studies of Neurodegenerative Disease: Part A

Edited by

MICHAEL J. HURLEY
Imperial College London, London, UK

ELSEVIER

AMSTERDAM • BOSTON • HEIDELBERG • LONDON
NEW YORK • OXFORD • PARIS • SAN DIEGO
SAN FRANCISCO • SINGAPORE • SYDNEY • TOKYO
Academic Press is an imprint of Elsevier

Academic Press is an imprint of Elsevier
225 Wyman Street, Waltham, MA 02451, USA
525 B Street, Suite 1800, San Diego, CA 92101-4495, USA
125 London Wall, London, EC2Y 5AS, UK
The Boulevard, Langford Lane, Kidlington, Oxford OX5 1GB, UK

First edition 2015

Notices

Knowledge and best practice in this field are constantly changing. As new research and experience broaden our understanding, changes in research methods, professional practices, or medical treatment may become necessary.

Practitioners and researchers must always rely on their own experience and knowledge in evaluating and using any information, methods, compounds, or experiments described herein. In using such information or methods they should be mindful of their own safety and the safety of others, including parties for whom they have a professional responsibility.

To the fullest extent of the law, neither the Publisher nor the authors, contributors, or editors, assume any liability for any injury and/or damage to persons or property as a matter of products liability, negligence or otherwise, or from any use or operation of any methods, products, instructions, or ideas contained in the material herein.

ISBN: 978-0-12-801480-6
ISSN: 0074-7742

For information on all Academic Press publications
visit our website at http://store.elsevier.com/

CONTENTS

Contributors ix
Preface xi

1. Alzheimer's Disease: Genomics and Beyond **1**

Fuhai Song, Guangchun Han, Zhouxian Bai, Xing Peng, Jiajia Wang, and Hongxing Lei

1. Introduction 2
2. GWASs on the Primary Phenotype of AD 3
3. GWASs on the Secondary Phenotype of AD 5
4. CNV Studies 7
5. WES and WGS Studies 7
6. Functional Genomics Studies 8
7. Brain Transcriptome Studies 8
8. Early-Onset AD 11
9. Epigenomics Studies in the Brain 11
10. AD, Brain Aging, and Longevity 13
11. AD and Relevant Diseases 13
12. Seeking Peripheral Biomarkers 14
13. Animal Studies of AD 15
14. iPSC Technology in AD Research 16
15. Integrating Multiomics Information for AD 17
16. Concluding Remarks 18

Acknowledgments 18
References 18

2. The Potential of Proteomics in Understanding Neurodegeneration **25**

Ramavati Pal, Jan Petter Larsen, and Simon Geir Moller

1. Introduction 26
2. Cellular Mechanisms of Neurodegeneration 27
3. Fundamentals of Proteomics 28
4. Application of Proteomics in Neurodegenerative Diseases 34
5. From Proteomics to Clinical Significance: Targeting Signaling Pathways in Neurodegenerative Diseases 47

6. Future Perspectives: Combinatorial Genomic, Metabolomics, and Proteomic Analysis 48
Acknowledgments 49
References 49

3. Proteomics Approach to Identify Biomarkers in Neurodegenerative Diseases 59

Annapurna Nayak, Gregory Salt, Sunil K. Verma, and Uday Kishore

1. Introduction 60
2. Proteomic Methods Utilized in Understanding Neurodegenerative Diseases 61
3. Alzheimer's Disease 63
4. AD Biomarkers 63
5. Parkinson's Disease 68
6. PD Biomarkers 69
7. Huntington's Disease 71
8. Prion Diseases 76
9. Prion Biomarkers 76
10. Amyotrophic Lateral Sclerosis 78
11. ALS Biomarkers 79
12. Conclusion 80

References 81

4. Uncovering Neurodegenerative Protein Modifications via Proteomic Profiling 87

Xavier Gallart-Palau, Aida Serra, and Siu Kwan Sze

1. Introduction 88
2. Nonenzymatic Degenerative Protein Modifications (DPMs) and Protein Dysfunctions 89
3. Analysis and Characterization of DPMs by Mass Spectrometry-Based Proteomics 92
4. Study of the Brain Proteome by Quantitative Proteomics 94
5. DPMs in Neurodegenerative Diseases 96
6. Protein Dysfunctions and Cognitive Impairment in Neurodegenerative Diseases 100
7. Concluding Remarks 106

Acknowledgements 107
References 107

5. Comparative Proteomics for the Evaluation of Protein Expression and Modifications in Neurodegenerative Diseases 117

Antonio Conti and Massimo Alessio

1. Introduction 118
2. Essential Neuroproteomics 119
3. Conclusions 144

References 145

Index 153

Contents of Recent Volumes 157

CONTRIBUTORS

Massimo Alessio
Proteome Biochemistry, Division of Genetics and Cell Biology, IRCCS San Raffaele Scientific Institute, Milano, Italy

Zhouxian Bai
CAS Key Laboratory of Genome Sciences and Information, Beijing Institute of Genomics, Chinese Academy of Sciences, and University of Chinese Academy of Sciences, Beijing, PR China

Antonio Conti
Proteome Biochemistry, Division of Genetics and Cell Biology, IRCCS San Raffaele Scientific Institute, Milano, Italy

Xavier Gallart-Palau
School of Biological Sciences, Nanyang Technological University, Singapore

Guangchun Han
CAS Key Laboratory of Genome Sciences and Information, Beijing Institute of Genomics, Chinese Academy of Sciences, Beijing, PR China

Uday Kishore
Centre for Infection, Immunity and Disease Mechanisms, College of Health and Life Sciences, Brunel University London, Uxbridge, United Kingdom

Jan Petter Larsen
Norwegian Center for Movement Disorders, Stavanger University Hospital, Stavanger, Norway

Hongxing Lei
CAS Key Laboratory of Genome Sciences and Information, Beijing Institute of Genomics, Chinese Academy of Sciences, and Center of Alzheimer's Disease, Beijing Institute for Brain Disorders, Beijing, PR China

Simon Geir Moller
Department of Biological Sciences, St. John's University, New York, USA, and Norwegian Center for Movement Disorders, Stavanger University Hospital, Stavanger, Norway

Annapurna Nayak
Centre for Infection, Immunity and Disease Mechanisms, College of Health and Life Sciences, Brunel University London, Uxbridge, United Kingdom, and Centre for Biotechnology and Bioinformatics, School of Life Sciences, Jawaharlal Nehru Institute for Advanced Studies, Secunderabad, Telangana, India

Ramavati Pal
Department of Biological Sciences, St. John's University, New York, USA

Xing Peng
CAS Key Laboratory of Genome Sciences and Information, Beijing Institute of Genomics, Chinese Academy of Sciences, and University of Chinese Academy of Sciences, Beijing, PR China

Gregory Salt
Centre for Infection, Immunity and Disease Mechanisms, College of Health and Life Sciences, Brunel University London, Uxbridge, United Kingdom

Aida Serra
School of Biological Sciences, Nanyang Technological University, Singapore

Fuhai Song
CAS Key Laboratory of Genome Sciences and Information, Beijing Institute of Genomics, Chinese Academy of Sciences, and University of Chinese Academy of Sciences, Beijing, PR China

Siu Kwan Sze
School of Biological Sciences, Nanyang Technological University, Singapore

Sunil K. Verma
Center for Cellular and Molecular Biology, Hyderabad, Telangana, India

Jiajia Wang
CAS Key Laboratory of Genome Sciences and Information, Beijing Institute of Genomics, Chinese Academy of Sciences, and University of Chinese Academy of Sciences, Beijing, PR China

PREFACE

If you can pass the anathematic word "omic" and choose to read this volume of the *International Review of Neurobiology*, then you will have done yourself a favor. Because nestled within the usual scientific prose is some truly outstanding and innovative work describing state-of-the-art omic investigations into the cause, diagnosis, and treatment of neurodegenerative disorders such as Alzheimer's disease, Parkinson's disease, Huntington's disease, amyotrophic lateral sclerosis, multiple sclerosis, and prion diseases.

The aim of the volume is to bring together diverse work that explores facets of neurodegenerative disease that are often overlooked by traditional hypothesis-driven research and to present such work in a readable format which appeals to researchers and students from a wide range of scientific backgrounds.

The reason for doing so is that neurodegenerative diseases ruin and frequently shorten the life of the patient and place immense strain upon their families and carers of the affected individual. This in turn has a high social and economic cost to society as a whole. Paradoxically, as advances in medicine successfully treat many illnesses and life expectancy increases, the prevalence of neurodegenerative disease is also rising, since the biggest risk factor for the most common neurodegenerative diseases is aging. Consequently, these costs will only get worse as the world population ages and the apparent advantages of a greater life expectancy are less desirable if the reality is that there is a high probability of developing a debilitating disease which renders the individual dependent on constant care. Yet, despite the vast quantities of research that has been conducted over many decades into neurodegenerative diseases, in the majority of cases the cause of neurodegeneration is unknown and treatments, if available, only tackle some symptoms of neurodegeneration and do nothing to slow the progression of the disease.

The chapters in this volume of *International Review of Neurobiology* showcase novel approaches to study the genome, proteome, and metabolome of samples from patients with neurodegenerative disease in the hope of finding clues regarding the causes of neurodegenerative disease, furthering our understanding of mechanisms of neurodegeneration, aiding earlier diagnosis through the detection of biomarkers, and identifying putative therapeutic targets that will allow successful treatment of neurodegenerative diseases or delay their onset.

Finally, I would like to thank the authors of the chapters in this volume for their work and hope that the readers find something within the volume that stimulates them to contribute toward further research into the cause, treatment, and cure of neurodegenerative disease in the future.

Michael J. Hurley
London, April 2015

CHAPTER ONE

Alzheimer's Disease: Genomics and Beyond

Fuhai Song[*,†], Guangchun Han[*], Zhouxian Bai[*,†], Xing Peng[*,†], Jiajia Wang[*,†], Hongxing Lei[*,‡,1]

[*]CAS Key Laboratory of Genome Sciences and Information, Beijing Institute of Genomics, Chinese Academy of Sciences, Beijing, PR China
[†]University of Chinese Academy of Sciences, Beijing, PR China
[‡]Center of Alzheimer's Disease, Beijing Institute for Brain Disorders, Beijing, PR China
[1]Corresponding author: e-mail address: leihx@big.ac.cn

Contents

1. Introduction 2
2. GWASs on the Primary Phenotype of AD 3
3. GWASs on the Secondary Phenotype of AD 5
4. CNV Studies 7
5. WES and WGS Studies 7
6. Functional Genomics Studies 8
7. Brain Transcriptome Studies 8
8. Early-Onset AD 11
9. Epigenomics Studies in the Brain 11
10. AD, Brain Aging, and Longevity 13
11. AD and Relevant Diseases 13
12. Seeking Peripheral Biomarkers 14
13. Animal Studies of AD 15
14. iPSC Technology in AD Research 16
15. Integrating Multiomics Information for AD 17
16. Concluding Remarks 18

Acknowledgments 18
References 18

Abstract

Alzheimer's disease (AD) is a major form of senile dementia. Despite the critical roles of Aβ and tau in AD pathology, drugs targeting Aβ or tau have so far reached limited success. The advent of genomic technologies has made it possible to gain a more complete picture regarding the molecular network underlying the disease progression which may lead to discoveries of novel treatment targets. In this review, we will discuss recent progresses in AD research focusing on genome, transcriptome, epigenome, and related subjects. Advancements have been made in the finding of novel genetic risk factors,

International Review of Neurobiology, Volume 121
ISSN 0074-7742
http://dx.doi.org/10.1016/bs.irn.2015.05.001

new hypothesis for disease mechanism, candidate biomarkers for early diagnosis, and potential drug targets. As an integration effort, we have curated relevant data in a database named AlzBase.

1. INTRODUCTION

Alzheimer's disease (AD) affects a large population in the senior community, likely 10 million in China alone (Han et al., 2014; Lei, 2010). The pathological hallmarks of AD include extracellular deposit of Aβ amyloid plaques derived from *APP* and intraneuronal neurofibrillary tangles (NFTs) from hyperphosphorylation of tau. Much of the efforts in AD research have been devoted to molecular pathways centered at Aβ or tau. The vast majority of novel treatment strategies are also targeting either Aβ or tau. Nevertheless, promising results from animal models have not translated well in human clinical trials (Callaway, 2012). Thus, revolutionary ideas outside of the hallmarks are desperately needed.

Technology developments in genomics have provided a variety of tools to investigate AD at the whole system level (Fig. 1). Earlier genetic linkage

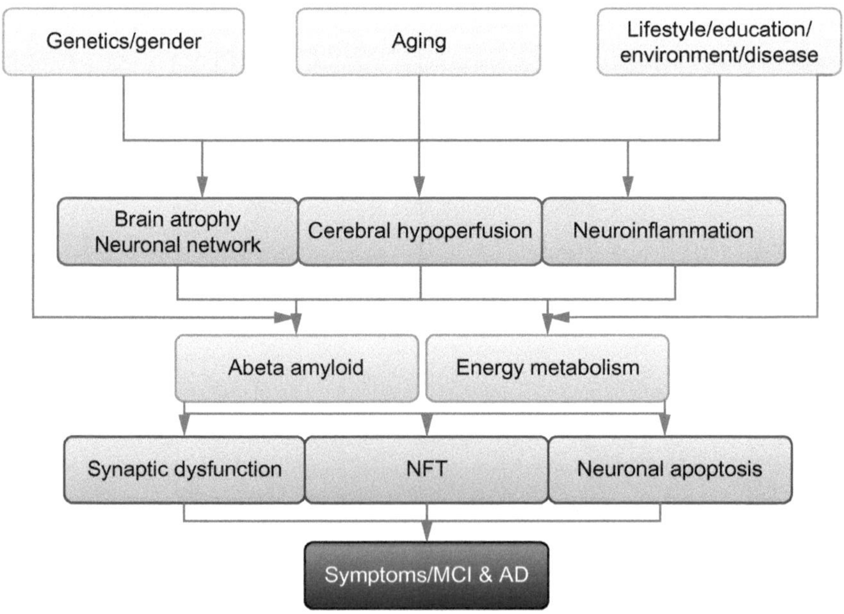

Figure 1 A framework of key factors and events in AD pathogenesis.

or association studies in both early-onset AD (EOAD) and late-onset AD (LOAD) have gradually evolved into genome-wide association studies (GWASs) and more recently developed into whole-exome sequencing (WES) and whole-genome sequencing (WGS; Bettens, Sleegers, & Van Broeckhoven, 2013; Guerreiro, Bras, & Hardy, 2013). Transcriptome studies are still in the process of transiting from microarray to RNA-seq (Sekar et al., 2015). Several types of epigenetic and epigenomic studies have been conducted on AD, including DNA methylation, histone modification, and microRNA (Lunnon & Mill, 2013). As a valuable supplement to the traditional studies on human subjects and animal models, induced pluripotency stem cell (iPSC) technology has been increasingly used to bridge the gap in translational medicine (Israel et al., 2012). In this review, we will survey representative investigations applying genomics technologies to the study of AD and relevant processes such as aging and neurological disorders.

2. GWASs ON THE PRIMARY PHENOTYPE OF AD

The quality of GWASs depends heavily on the sample size. In the most recent large-scale GWAS of AD (Lambert et al., 2013), over 74K samples were included in the meta-analysis (all of European ancestry), including ~54K samples in stage 1 (previously published datasets) and ~20K samples in stage 2 (new datasets). Over 7M imputed single-nucleotide polymorphisms (SNPs) were used in the meta-analysis of stage 1 datasets. In stage 2, only SNPs showing moderate significance in stage 1 ($P < 1 \times 10^{-3}$) were genotyped (~11K SNPs). In the meta-analysis of stage 1 datasets, the significance of *APOE* and nine other previously identified genes was confirmed (Table 1), including *CR1*, *BIN1*, *CLU*, *PICALM*, *ABCA7*, *CD2AP*, *EPHA1*, *MS4A4E*, and *CD33*. Additionally, five new loci reached genome-wide significance in the discovery datasets, namely, *SORL1*, *PTK2B*, *DSG2*, *HLA-DRB5*, and *SLC24A4*. When combining the discovery and replication datasets, seven new loci reached genome-wide significance, including *INPP5D*, *MEF2C*, *NME8*, *FERMT2*, *ZCWPW1*, *CELF1*, and *CASS4*. Among these 22 loci, *CD33* and *DSG2* did not reach genome-wide significance in the combined datasets. These genes are involved in immune response, lipid metabolism, synaptic function, and Aβ/tau-related pathways.

Besides testing the significance at the SNP level, gene-level test may reveal new genes affecting AD susceptibility. In a follow-up work of the above-described study (Escott-Price et al., 2014), gene-level analysis

Table 1 Key Genes from Genetic Studies of Late-Onset AD

Genes	Full Name	Relevant Functions
CR1	Complement component (3b/4b) receptor 1 (Knops blood group)	Complement pathway; clearance of amyloid
BIN1	Bridging integrator 1	Synaptic vesicle endocytosis
CLU	Clusterin	Stress response; lipid metabolism
PICALM	Phosphatidylinositol-binding clathrin assembly protein	Synaptic vesicle; lipid metabolism
ABCA7	ATP-binding cassette, subfamily A (ABC1), member 7	Lipid metabolism; immune
CD2AP	CD2-associated protein	Regulation of actin cytoskeleton; endocytosis
EPHA1	EPH receptor A1	Signaling in nervous system development; immune
MS4A4E	Membrane-spanning four domains, subfamily A, member 4E	Immune; signaling
CD33	CD33 molecule	Immune; signaling
SORL1	Sortilin-related receptor, L(DLR class) A repeats containing	Endocytosis and sorting
PTK2B	Protein tyrosine kinase 2 beta	Synaptic transmission
DSG2	Desmoglein 2	Cell adhesion
HLA-DRB5	Major histocompatibility complex, class II, DR beta 5	Immune response
SLC24A4	Solute carrier family 24 (sodium/potassium/calcium exchanger), member 4	Ion transport
INPP5D	Inositol polyphosphate-5-phosphatase, 145 kDa	Blood cell; immune
MEF2C	Myocyte enhancer factor 2C	Transcription regulation; brain development
NME8	NME/NM23 family member 8	Cytoskeleton
FERMT2	Fermitin family member 2	Cell–matrix adhesion
ZCWPW1	Zinc finger, CW type with PWWP domain 1	Epigenetic regulation; blood trait

Table 1 Key Genes from Genetic Studies of Late-Onset AD—cont'd

Genes	Full Name	Relevant Functions
CELF1	CUGBP, Elav-like family member 1	mRNA processing; cytoskeleton
CASS4	Cas scaffolding protein family member 4	Cell adhesion; signaling
TP53INP1	Tumor protein p53-inducible nuclear protein 1	Autophagic cell death
IGHV1-67	Immunoglobulin heavy variable 1–67 (pseudogene)	Immune response
PDE1A	Phosphodiesterase 1A, calmodulin dependent	Calcium signaling
RYR3	Ryanodine receptor 3	Calcium signaling

identified two genes with no previous reports, *TP53INP1* and *IGHV1-67*. The three most significant SNPs in *TP53INP1* represented at least two partially independent signals, while the two most significant SNPs in *IGHV1-67* represented the same signal. As for the gene function, *TP53INP1* is involved in apoptosis and *IGHV1-67* is involved in immune response.

We have adopted a different strategy in our reanalysis of public GWAS datasets. We considered up to four SNPs when evaluating intragenic epistasis using a generalized multidimension reduction algorithm (Sun et al., 2014). When considering 4-order intragenic epistasis, 10 genes were found to have consistent strong signal across the four publicly available GWAS datasets for AD. Among these 10 genes, *PDE1A* and *RYR3* may deserve further attention due to their involvement in calcium signaling which is known to be dysregulated in AD brains (Demuro, Parker, & Stutzmann, 2010).

3. GWASs ON THE SECONDARY PHENOTYPE OF AD

Age at onset (AAO) is a major secondary phenotype studied on AD. In the first GWAS of AAO of AD, several SNPs at the *APOE* region reached genome-wide significance, while *DCSH2* was a newly found risk gene affecting AAO of AD (Kamboh et al., 2012). In a study focused on the top 10 risk loci, it was found that the risk allele of *PICALM* may lead to slightly earlier onset of AD, while the strong effect of *APOE* e4 allele on AAO was confirmed (Thambisetty, An, & Tanaka, 2013). In another study focused on the *TOMM40* intron 6 poly-T length, no association was found with the AAO in LOAD, but it may affect the neuropathology in AD (Li et al., 2013).

Cognition and cognitive decline in AD and general population is another area of interest. In a longitudinal study of nondemented elderlies between age 79 and 87, *APOE* e4 was found to be associated with cognitive decline regarding verbal memory and abstract reasoning (Schiepers et al., 2012). In a GWAS on 3511 unrelated adults, *APOE* e4 was found to be the only independent genome-wide significant signal associated with cognitive aging (Davies et al., 2014). In another longitudinal study over 54 months on healthy elderlies, the effect of *APOE* e4 on the rate of cognitive decline was found to be dependent on the presence of amyloid in the brain (Lim et al., 2014). In two studies on the top 10 risk loci for AD, collective effect of the risk loci on cognitive aging independent of *APOE* was found to be marginal (Carrasquillo et al., 2015; Verhaaren et al., 2013). Genome-wide polygenic risk score was also found to exert no effect on cognitive aging in a study on over 3000 nondemented old people (Harris et al., 2013). In two separate GWASs, *SPON1* was found to be the only gene affecting both cognitive decline (Sherva et al., 2014) and brain wiring (Jahanshad et al., 2013).

Another relevant phenotype is hippocampal volume (HV) or cortical thickness. An early GWAS on brain atrophy identified *ZNF292* as a novel genetic factor affecting entorhinal cortical thickness (Furney et al., 2011). In two large-scale studies on the general population, several novel loci were found to be associated with HV, including *TESC*, *MFRB3-WIF1*, and *HRK-FBXW8* (Bis et al., 2012; Stein et al., 2012). In a study on AD cases and relevant controls, four genes were found to be associated with HV, including *APOE*, *F5/SELP*, *LHFP*, and *GCFC2* (Melville et al., 2012). In a related study, two novel genes (*CARD10* and *PARP1*) were found to affect hippocampal degeneration and HV in *APOE* e3/e3 individuals (Nho et al., 2013).

Amyloid plaque burden is also of intense interest in the AD research community. In the first GWAS of amyloid plaque in 555 live patients, *APOE* and *BCHE* reached genome-wide significance with independent as well as additive effect on plaque load (Ramanan et al., 2014). In a study combining the target gene and GWAS on both deceased subjects and live patients, four target genes (*APOE*, *CR1*, *ABCA7*, and *CD2AP*) were associated with plaque burden. In addition, the GWAS revealed a novel risk locus near *APP* (Shulman et al., 2013).

4. CNV STUDIES

Copy number variation (CNV) is suggested to be a part of the missing heritability in AD. A small-scale genome-wide CNV studies were conducted on ~700 subjects. *CHRNA7* was found to be one of the genes with suggestive significance (Heinzen et al., 2010). A similar study on Caribbean Hispanics revealed nominal association at Chr15q11.2 (Ghani et al., 2012) encompassing five genes. In a study on the AD neuroimaging initiative samples, significant higher burden of CNV-region deletions was found in mild cognitive impairment (MCI) and AD cases compared to controls (Guffanti et al., 2013). In two separate studies on the effect of CNV on the AAO of AD, CNVs in *CPNE4* and *CHRFAM7A* were found to be potential risk for AD (Szigeti et al., 2014, 2013).

5. WES AND WGS STUDIES

The genome-wide significant SNPs from GWASs can only explain a small portion of the AD heritability. Some of the genetic risks are hidden in the rare coding variants which are not assessed in GWASs. The fast lowering of sequencing cost made it possible to conduct WES or even WGS studies on AD. One such study focused on *TREM2*, a gene known to be associated with an autosomal recessive form of early-onset dementia (Guerreiro et al., 2012). The study started with sequencing *TREM2* in ~2000 samples (approximately half AD patients and half controls). It was found that exon 2 of this gene had significantly more variants in patients compared to controls. The most significant variant rs75932628 (R47H mutation) was confirmed to be significant in GWAS datasets through SNP imputation. The significance of this variant was further validated through direct genotyping on additional ~60K samples. The functional relevance of *TREM2* to AD was demonstrated by the elevated expression in a mouse model of AD compared to control.

Another study started with WES of 14 selected LOAD families (Cruchaga et al., 2014). The WES combined with genotyping revealed only one rare variant (*PLD3*, V232M) segregating the disease status in two independent families. The significance of this variant was confirmed in independent datasets with ~11K samples. Furthermore, *PLD3* gene was resequenced in ~4400 European samples and ~300 African-American samples and the gene-level significance was validated. To evaluate the function of *PLD3* in AD, overexpression and knockdown experiments were performed. It was found that *PLD3* level was strongly correlated with the level of extracellular Aβ42 and Aβ40.

6. FUNCTIONAL GENOMICS STUDIES

Compared to the well-accepted results on the odds ratio of the above-described genes from GWAS or WES studies, little consensus has been reached regarding the functional relevance of those genes and more specifically the SNP variations. Based on the examination of 73 cases and 39 controls, it was found that the expression level for several of the top 10 genes demonstrated association with various clinical assessment of AD including clinical dementia rating score, disease progression rate, disease status, and Braak stage (Karch et al., 2012). However, none of the examined SNPs showed association with gene expression level, although a few suggestive associations with secondary phenotypes were observed. In a study focused on the loss-of-function effect on Aβ production, it was shown that most of the top LOAD genes did not alter Aβ production except for *CLU* and *CD2AP* (Bali, Gheinani, Zurbriggen, & Rajendran, 2012). More importantly, none of the top LOAD genes affected Aβ42/40 ratio in the RNAi experiments. In an expression quantitative trait loci (eQTL) study, none of the top loci showed a regulatory role on gene expression (Holton et al., 2013). In the meantime, the brain-wide regional difference in gene expression showed no correlation with the difference in regional vulnerability.

7. BRAIN TRANSCRIPTOME STUDIES

Brain transcriptome as an endophenotype provides the most comprehensive information regarding the dynamic change of molecular network during disease development. Through gene expression profiling from various aspects of AD development, the systematic picture is becoming increasingly clear. In a pioneering study, hippocampus tissues from 31 subjects at four stages of AD development were examined, including normal elderly controls and AD patients at the incipient, moderate, and severe stages (Blalock et al., 2004). In addition to transcriptome itself, the correlation between gene expression and mini-mental state examination or NFT scores was evaluated. The major finding was the upregulation of transcriptional and tumor suppressor response at the incipient stage, accompanied by the downregulation of protein folding/metabolism/transport and energy metabolism.

Detailed Braak staging was adopted in a more recent work on the prefrontal cortex (PFC). Brain samples from Braak stage 0–6 were examined. Significant genes were clustered based on the temporal profiles of gene expression. Among the four important gene clusters, two displayed up-down feature and the other two displayed down-up feature, both with the turning point at Braak stage 2. Parallel measurement of intraneuronal Aβ showed decreased Aβ level at Braak stages 4–6. It was suggested that these two events were connected and the change of gene expression at the early stage may be a coping mechanism against the increase of soluble Aβ level in the PFC. Further dissection of *APOE* genotype revealed the potential role of *APOE* in accelerating gene expression change.

The development of AD pathology is not homogeneous across the brain. To examine the brain region-specific response to AD development, six relevant brain regions were selected for in-depth evaluation in a well-designed study employing laser-capture microdissection technology (Liang et al., 2008). For direct comparison with healthy elderly controls, only nontangle-bearing neurons were selected for gene expression profiling. The selected brain regions include hippocampus, entorhinal cortex, middle temporal gyrus, posterior cingulate cortex, superior frontal gyrus, and primary visual cortex (VCX). The brain-wide comparison revealed that different brain regions had distinctive response to AD pathology using the healthy controls as reference.

Due to the high quality of this dataset, we conducted reanalysis using a protein–protein interaction (PPI) network approach (Liang et al., 2012). First, we identified the perturbed network in each of the six brain regions. Subsequently, we found that there was high level of overlap among the perturbed networks in the six brain regions. This indicated that the molecular response to the AD pathology was similar in different brain regions, although the degree of perturbation may vary from one brain region to another. We further selected the top 10% most connected genes from each of the six perturbed networks. Interestingly, these 136 hub genes themselves formed a highly connected network. We demonstrated that the hub network was highly relevant to AD mechanism using several independent validations. Drug target enrichment analysis on the hub network further supported the potential application of the network in drug discovery for AD.

Small sample size is the major limitation for most brain transcriptome studies including AD. The first large-scale AD transcriptome study employed 364 brain samples (approximately half for cases and half for controls; Webster et al., 2009). In addition to the case–control comparison, the study focused on the genetic control of gene expression by concurrent genotyping and gene expression profiling of the samples. The analysis revealed that the expression of 9% of the transcripts may be controlled by SNPs and 5% of the transcripts had LOAD-specific SNP-transcript correlation.

In a recent large-scale brain transcriptome study of AD, over 1600 brain tissues from three brain regions were examined (Zhang et al., 2013). The brain regions included PFC, cerebellum (CB), and primary VCX. Not surprisingly, the PFC region displayed much higher level of perturbation than the other two brain regions. Through integrated network analysis of genome and transcriptome, the immune and microglia module was found to have a significant role in AD pathogenesis. *TYROBP* was identified as a driver gene of this module and the causal relationship was validated by overexpressing *TYROBP* and measuring downstream effect.

In order to investigate gene dysregulation pattern during AD progression, we selected four datasets representing four stages of AD development. We found that downregulation of energy metabolism occurred at the early stage where dementia symptom was not yet manifested (Sun, Feng, Liang, Duan, & Lei, 2012). Downregulation of energy metabolism has been considered as a consequence of mitochondrial damage due to oxidative stress. However, the downregulation already occurred in tangle-free neurons at the nondementia stage where no apparent signals for oxidative stress were detected. Therefore, we proposed that the downregulation of energy metabolism is a protective response of neurons under the environment with deficiency of energy supply. Energy conservation might be the only strategy for the neurons to survive.

In order to examine the consistency among the brain transcriptome studies, we have also conducted a comprehensive survey on all of the publicly available datasets for AD transcriptome (Feng et al., 2014). We first demonstrated that gene dysregulation in the AD brains was highly reproducible. Then, we selected 100 genes representing the gene dysregulation in the AD brains. We validated the robust dysregulation of this set of genes with independent datasets. Among these 100 genes, 12 genes displayed consistent up- or downregulation along the trajectory of AD development, suggesting higher relevance to the disease progression. In addition, we also found an interesting scenario of gene expression cushion in the less vulnerable brain regions.

8. EARLY-ONSET AD

While most transcriptome studies focused on LOAD, one study examined monogenic and sporadic EOAD (Antonell et al., 2013). When compared to the control group, ~3000 dysregulated genes were found for both types of EOAD. The functional categories were also similar, including calcium signaling, neuroactive ligand–receptor interaction, axon guidance, and long-term potentiation. Direct comparison between sporadic and monogenic EOAD revealed little difference. It was suggested that these two types of EOAD converge to a similar final stage of the disease. We shall note that the dysregulated functional categories are also similar to those in LOAD, although the affected brain regions may be different.

Several studies have been conducted on the genetics of EOAD. In an earlier study on the AAO of familiar AD carrying *PSEN2* mutations, three loci were found to have effect on AAO, including 1q23.3, 17p13.2, and 7q33 (Marchani et al., 2010). In a study on familiar AD subjects in Caribbean Hispanics, 12 mutations were found in the 5 selected genes (*APP*, *PSEN1*, *PSEN2*, *MAPT*, and *GRN*), among which 4 were novel mutations (Lee et al., 2014). In a genome-wide rare CNV study of 261 families with EOAD, 10 novel private CNVs were found in gene-rich regions (Hooli et al., 2014).

9. EPIGENOMICS STUDIES IN THE BRAIN

Much of the epigenomics studies of AD are focused on DNA methylation (5-mC) and hydroxymethylation (5-hmC). However, inconsistent results have been reported in the literature. In an earlier study by Rogers and coworkers (Mastroeni et al., 2010), two markers for DNA methylation and eight key methylation factors were examined in entorhinal cortex layer II. Lower level for all 10 factors was observed in AD patients compared to controls. A more recent study by van den Hove and coworkers demonstrated robust decrease of 5-mC and 5-hmC in the hippocampus of AD patients and similarly in the AD twin examined in the same study (Chouliaras et al., 2013). In addition, the methylation level showed significant negative correlation with amyloid plaque load. Another study by Lunnon and coworkers reported significant decrease of 5-hmC level in both entorhinal cortex and CB of AD patients (Condliffe et al., 2014).

On the other hand, the study by Balazs and coworkers reported no significant change of global levels of 5-mC and 5-hmC in the entorhinal cortex of AD patients (Lashley et al., 2014). A study by Lovell and coworkers reported increase of 5-mC and 5-hmC in hippocampus/parahippocampal gyrus of preclinical and late-stage AD patients (Bradley-Whitman & Lovell, 2013). Recently, the study by Dragunow and coworkers demonstrated global increase of 5-mC and 5-hmC in middle frontal gyrus and middle temporal gyrus of AD patients (Coppieters et al., 2014). An earlier study by Rapoport and coworkers also reported global hypermethylation in the frontal cortex of AD patients (Rao, Keleshian, Klein, & Rapoport, 2012).

The study by Rozek and coworkers focused on the locus-specific change of methylation in frontal cortex by methylation array experiment (Bakulski et al., 2012). A total of 948 CpG sites were reported to be potentially associated with AD phenotype, among which a CpG site at the promoter region of *TMEM59* demonstrated the highest discrimination power.

Several studies are focused on the deregulation of microRNA levels in the brain and the potential regulatory mechanism. The most consistent conclusion is the deregulation of miR-132 family in AD patients. Lau et al. found 35 deregulated microRNAs in the hippocampus and 41 deregulated microRNAs in the PFC of AD patients (Lau et al., 2013). The strongly deregulated miR-132-3p in both brain regions was further validated and the deregulation started at Braak stage 3. Functionally, miR-132-3p may regulate tau pathway through *FOXO1a*. In another study by Hebert et al. using sequencing technology, miR-132-3p was also found to be significantly decreased in the temporal cortex of AD patients (Hebert, Wang, Zhu, & Nelson, 2013). In yet another study, Wong et al. found downregulation of miR-132 and miR-212 in hippocampus and temporal cortex of AD patients (Wong et al., 2013). They further demonstrated that miR-132/212 may regulate neuronal apoptosis through AKT signaling pathway.

Deregulation of other microRNAs has also been reported. An earlier study by Hebert et al. demonstrated lower miR-29 cluster in AD patients, which may lead to the increase of *BACE1* and Aβ level (Hebert et al., 2008). In a study by Wang et al. with more disease stages, it was demonstrated that the decrease of miR-107 was found in early AD which may also regulate Aβ level through the increase of *BACE1* (Wang et al., 2008). In a study by Cui et al. on the hippocampus and neocortex, increased level of miR-146a was found in AD which may together with *NFkB* play a critical role in regulating the *IRAK* level (Cui, Li, Zhao, Bhattacharjee, & Lukiw, 2010).

It is encouraging to see consistent observations on some of the studies on DNA methylation and microRNA in the brain. However, due to the limited number of such studies and small sample size in those studies, caution shall be taken when interpreting the results.

10. AD, BRAIN AGING, AND LONGEVITY

Aging is the primary risk factor for AD; thus, studies on brain aging are highly relevant to AD. An earlier study focused on frontal cortex spanning age 26–106. Downregulation of synaptic transmission and mitochondrial function was observed starting at 40, followed by the upregulation of stress response and DNA repair (Lu et al., 2004). In addition, DNA damage was found in the promoter regions of the downregulated genes. Another study assessed four distinctive brain regions at age 20–99. Superior frontal gyrus was found to be the most significantly dysregulated brain region and the most prominent change was observed from 60 to 70 (Berchtold et al., 2008). In addition, clear gender difference was found during the aging process, including the total number of affected genes and the exact nature of the affected functional categories.

Genetic studies have also been conducted on aging and longevity. In an earlier meta-analysis of aging-related GWAS datasets, no genome-wide significant SNPs were found, although 22 SNPs demonstrated moderate prediction power on the risk of death or event-free survival (Walter et al., 2011). A study on people with exceptional longevity revealed a panel of genetic signature with 281 SNPs for longevity (Sebastiani et al., 2012). In a genome-wide linkage study of human longevity, *APOE* was identified as a major contributor to longevity (Beekman et al., 2013). Three additional loci were found in 19p13.3-11, 17q21-22, and 14q11.2. An epigenomic scan revealed 490 differentially methylated regions correlated with chronological age (Bell et al., 2012). As a reference source, LongevityMap has curated genes associated with aging and longevity from genetic studies (Budovsky et al., 2013).

11. AD AND RELEVANT DISEASES

Due to the similarities in symptoms and disease pathology, investigations have been conducted to unveil the common molecular mechanism between AD and related diseases. A study on AD and age-related macular degeneration revealed common genetic risks in *ABCA7* and *ZCWPW1* (Logue et al., 2014). In a study between schizophrenia (SZ) and general cognitive ability, a polygenic risk score for SZ was found to affect cognitive ability (Lencz et al., 2014). Results from other investigations, however, mostly turned out to be negative. A recent combined GWAS revealed no significant SNP conferring risk for both AD and Parkinson's disease (PD; Moskvina et al., 2013). A study on Korean population also showed no significant effect of the AD top loci on the susceptibility of PD (Chung et al., 2013). An investigation on the genetic link between type 2 diabetes (T2D) and AD also revealed no significant effect for the T2D top loci on AD susceptibility (Proitsi et al., 2014).

12. SEEKING PERIPHERAL BIOMARKERS

Due to the difficulty in accessing the brain with noninvasive approaches, peripheral biomarkers have been pursued for the early diagnosis of AD. An earlier study on blood transcriptome revealed lower expression of cytoskeletal integrity and DNA repair among other functions (Maes et al., 2007). A later study identified a 170-probe signature with high sensitivity and specificity distinguishing AD and elderly controls. Most of the signature genes were involved in immune function and lipid metabolism (Fehlbaum-Beurdeley et al., 2010). In another study, a 96-gene set was derived and displayed good sensitivity and specificity (Booij et al., 2011; Rye et al., 2011). The major functions included nucleic acid metabolism, protein metabolism, apoptosis, and cell cycle. In a study on both AD and MCI, lipid metabolism was among the dysregulated functions in AD and *ABCB1* displayed good correlation with disease progression (Chen et al., 2011). In another study with similar design, increased expression of immune response and decreased expression of mitochondria were identified in AD patients as well as in the early stage of AD (MCI; Lunnon et al., 2012).

Since peripheral blood circulates through the whole body, the specificity is a major concern when deriving biomarkers from the blood. We collected blood transcriptome datasets from investigations on various diseases including neurological disorders, cancer, diabetes, and infectious diseases. We found that the concurrent downregulation of core metabolism and upregulation of environmental response is a unique feature for AD (Han et al., 2013). We also validated this finding using a small Chinese cohort of AD, MCI, and elderly control. In addition, we found that the degree of perturbation in oxidative phosphorylation went up as the disease progressed. Some of the critical genes were further discussed in our collaborative investigation using samples from the United States (Bai et al., 2014).

In addition to the expression level of genes, microRNA expression in the peripheral blood has also been investigated. In a study focused on MCI, miR-132 and miR-134 families were found to have discrimination power (Sheinerman et al., 2012); further validation using large independent samples demonstrated 96% and 87% overall accuracy for these two microRNA families, respectively (Sheinerman, Tsivinsky, Abdullah, Crawford, & Umansky, 2013). In a study focused on AD, six microRNAs were found to have good discrimination power, among which miR-342-3p displayed the best performance (Tan et al., 2014). In another similar study, seven microRNAs including let-7d-5p were able to separate AD from normal controls with high accuracy (Kumar et al., 2013). In another study with both AD and MCI, a panel of 12 microRNAs was found to have good accuracy separating AD from controls as well as other neurological disorders (Leidinger et al., 2013). Despite the progress, one of the main concerns in this field is the inconsistency among independent studies which shall be taken into account in the future.

13. ANIMAL STUDIES OF AD

Animal models especially mouse models have been widely used in the investigation of the molecular mechanism of AD. In one such study on the 5XFAD mouse model, cardiovascular disease-related genes were found to be downregulated in the frontal cortex, while mitochondrial genes were downregulated in CB (Kim, Moon, Yu, Mook-Jung, & Kim, 2012). In another study, two types of transgenic mice (5XFAD and Tg4-42) were both compared with wild-type mice, and a set of 36 common differentially expressed genes were found which may underlie the common molecular pathways leading to the plaque pathology (Bouter et al., 2014). In a rare study on *Microcebus murinus*, both old animals and AD-like animals were compared to young animals (Abdel Rassoul et al., 2010). Opposite trend of gene dysregulation was found in certain functional categories especially metabolic pathways, which was suggested as a compensatory mechanism.

Epigenetic studies have also been conducted on mouse models. In a DNA methylation study, differentially methylated regions were found in two types of mouse models for AD (Sanchez-Mut et al., 2013). The hypermethylation of three genes (*TBXA2R*, *SORBS3*, and *SPTBN4*) was consistent with the downregulation of the three genes in human AD. In a study on both aging and AD, global decrease of 5-mC was observed during the aging process, while increase of 5-hmC was observed in 3xTgAD mouse compared to wild-type control (Cadena-del-Castillo et al., 2014). In a study on histone modification, the level of H3/H4 acetylation started to increase at 4 months of age in 3xTgAD mouse, which is a distinctive feature compared to wild-type mice (Walker, LaFerla, Oddo, & Brewer, 2013). In a microRNA study, the upregulation of microRNA in 3xTgAD mouse converged to the regulation of synaptic function, and miR-325 was identified as a key player in both AD pathology development and response to environmental enrichment (Barak et al., 2015). In another microRNA study, upregulation of miR-181 was found in 12-month-old 3xTgAD mouse (Rodriguez-Ortiz, Baglietto-Vargas, Martinez-Coria, LaFerla, & Kitazawa, 2014). Concurrent decrease of *c-Fos* and *SIRT1* was found which might be regulated by miR-181. The level of miR-206 was found to be higher in an embryonic *APP/PS1* transgenic mouse covering several tissues including hippocampal tissue, cerebrospinal fluid, and plasma (Tian, Cao, & Zhang, 2014). This increase of miR-206 was proposed as a key regulator of *BDNF* which was lower in the transgenic model. In another study, miR-342-5p was found to be upregulated in three transgenic models including the *APP/PS1* mice. The upregulation was linked to several genes including *ankG* which is involved in axon guidance (Sun, Wu, Gu, & Zhang, 2014).

14. iPSC TECHNOLOGY IN AD RESEARCH

The iPSC technology has been applied to the understanding of the functional consequence of critical mutations using isogenic human stem cell. Several works have been conducted on the *PSEN1* mutations. In one such study, *PS1* mutation was found to affect the γ-secretase function but not other functions of *PS1*. Therefore, the mutational effect is a mixture of loss of and gain of functions (Woodruff et al., 2013). In another study, Aβ42/Aβ40 ratio was found to be higher in the *PS1* mutant NPC. The study further revealed 14 genes dysregulated in the *PS1* mutant NPC, among which 5 genes showed consistent feature in LOAD (Sproul et al., 2014). In another study on drug response, higher Aβ42/Aβ40 ratio was also observed in the mutant *PS1* neurons compared to the wild type. More importantly, a novel drug response signature was observed when treating the cell culture with a GSM, suggesting the value of using human-derived neurons (Liu et al., 2014). A study focused on the protective effect of A673T of *APP* from an earlier finding. The authors demonstrated that the mutation led to reduced cleavage of *APP* by *BACE1* as well as reduced aggregation of Aβ42 (Maloney et al., 2014).

15. INTEGRATING MULTIOMICS INFORMATION FOR AD

The fast accumulation of data from systematic studies of AD has created a daunting task for data integration. Different types of omics studies are not isolated; the integration requires great effort in order to reach the clearer picture of the jigsaw puzzle. We have attempted to bridge the gap between genetic and transcriptome studies by evaluating the transcriptional regulation in different chromosome regions. We found that chr19p displayed the most significant dysregulation in three stages of AD development (Wang et al., 2014). This was especially interesting considering the fact that a strong signal was observed in chr19p in previous genetic studies. In addition, *APOE*, the most consistent genetic risk for LOAD, resides on the same chromosome. Therefore, we proposed several critical genes such as *C3* and *KANK2* on chr19p.

Recently, we collected omics data related to AD and constructed a database named AlzBase as a reference for gene dysregulation in AD and relevant processes (Fig. 2). The core data were the gene dysregulation in brain

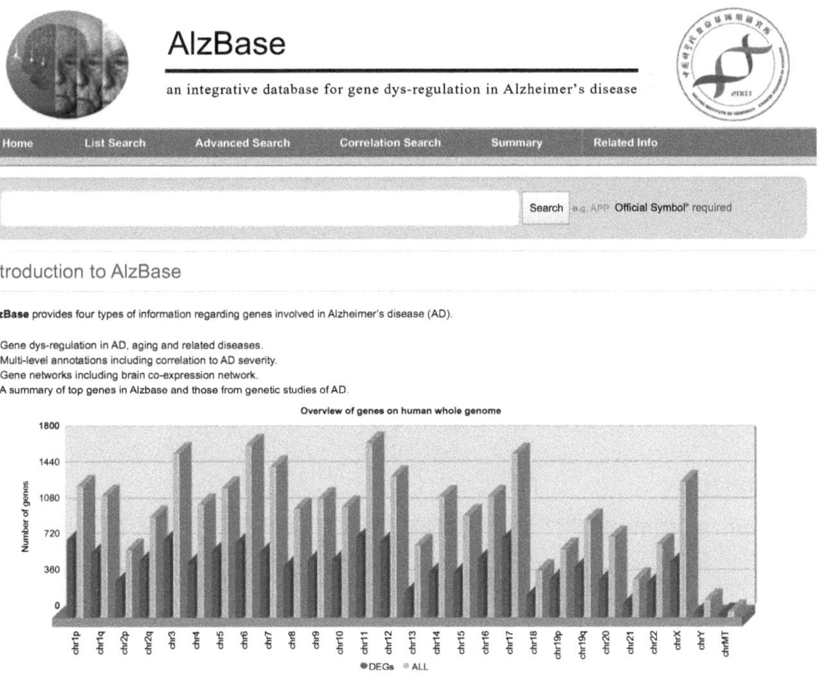

Figure 2 The homepage of AlzBase—an integrative genomics database for AD.

transcriptome datasets for AD. Genes correlated with AD progression were also curated. This was further extended to gene dysregulation in brain aging, other neurological diseases, and T2D. Other data included results from blood transcriptome studies, GWASs, eQTL studies, and microRNAs and drugs relevant to AD. To further illustrate the relationship among the critical genes, we provided several types of networks including coexpression network, PPI network, mutual information network, and cocitation network. The first version of AlzBase (http://alz.big.ac.cn/alzBase/) is available online and we plan to update the database to keep up with the development in this field (Bai et al., 2015).

16. CONCLUDING REMARKS

The field of AD research has advanced tremendously while embracing the fast-growing omics technologies. The development of sequencing technology is making personal genome increasingly accessible to the general public as well as patients with specific diseases. It will also make the dynamic monitoring of the blood transcriptome a more realistic goal in the near future. The development of iPSC technology including 3D neuron culture will make the disease model more human-like to avoid costly failures in traditional drug development employing mouse models. On another note, neuroimaging, which is not covered in this review, will certainly play an important role in visualizing the molecular details of the dysfunctional network. With the increasing use of these technologies, we expect to see exciting advancements in the field of AD research.

ACKNOWLEDGMENTS

This work was supported by the grant from the National Basic Research Program of China (973 Program; Grant No. 2014CB964901) and National High Technology Program of China (863 Program; Grant No. 2015AA020108) awarded to H.L. from the Ministry of Science and Technology of China.

Competing Interests: The authors declare that there are no conflicts of interest.

REFERENCES

Abdel Rassoul, R., Alves, S., Pantesco, V., De Vos, J., Michel, B., Perret, M., et al. (2010). Distinct transcriptome expression of the temporal cortex of the primate Microcebus murinus during brain aging versus Alzheimer's disease-like pathology. *PLoS One*, *5*(9), e12770.

Antonell, A., Llado, A., Altirriba, J., Botta-Orfila, T., Balasa, M., Fernandez, M., et al. (2013). A preliminary study of the whole-genome expression profile of sporadic and monogenic early-onset Alzheimer's disease. *Neurobiology of Aging*, *34*, 1772–1778.

Bai, Z., Han, G., Xie, B., Wang, J., Song, F., Peng, X., et al. (2015). AlzBase: An integrative database for gene dysregulation in Alzheimer's disease. *Molecular Neurobiology*. http://dx.doi.org/10.1007/s12035-014-9011-3.

Bai, Z., Stamova, B., Xu, H., Ander, B. P., Wang, J., Jickling, G. C., et al. (2014). Distinctive RNA expression profiles in blood associated with Alzheimer disease after accounting for white matter hyperintensities. *Alzheimer Disease and Associated Disorders*, *28*, 226–233.

Bakulski, K. M., Dolinoy, D. C., Sartor, M. A., Paulson, H. L., Konen, J. R., Lieberman, A. P., et al. (2012). Genome-wide DNA methylation differences between late-onset Alzheimer's disease and cognitively normal controls in human frontal cortex. *Journal of Alzheimer's Disease*, *29*, 571–588.

Bali, J., Gheinani, A. H., Zurbriggen, S., & Rajendran, L. (2012). Role of genes linked to sporadic Alzheimer's disease risk in the production of beta-amyloid peptides. *Proceedings of the National Academy of Sciences of the United States of America*, *109*, 15307–15311.

Barak, B., Shvarts-Serebro, I., Modai, S., Gilam, A., Okun, E., Michaelson, D. M., et al. (2015). Opposing actions of environmental enrichment and Alzheimer's disease on the expression of hippocampal microRNAs in mouse models. *Translational Psychiatry*, *3*, e304.

Beekman, M., Blanche, H., Perola, M., Hervonen, A., Bezrukov, V., Sikora, E., et al. (2013). Genome-wide linkage analysis for human longevity: Genetics of Healthy Aging Study. *Aging Cell*, *12*, 184–193.

Bell, J. T., Tsai, P. C., Yang, T. P., Pidsley, R., Nisbet, J., Glass, D., et al. (2012). Epigenome-wide scans identify differentially methylated regions for age and age-related phenotypes in a healthy ageing population. *PLoS Genetics*, *8*, e1002629.

Berchtold, N. C., Cribbs, D. H., Coleman, P. D., Rogers, J., Head, E., Kim, R., et al. (2008). Gene expression changes in the course of normal brain aging are sexually dimorphic. *Proceedings of the National Academy of Sciences of the United States of America*, *105*, 15605–15610.

Bettens, K., Sleegers, K., & Van Broeckhoven, C. (2013). Genetic insights in Alzheimer's disease. *Lancet Neurology*, *12*, 92–104.

Bis, J. C., DeCarli, C., Smith, A. V., van der Lijn, F., Crivello, F., Fornage, M., et al. (2012). Common variants at 12q14 and 12q24 are associated with hippocampal volume. *Nature Genetics*, *44*, 545–551.

Blalock, E. M., Geddes, J. W., Chen, K. C., Porter, N. M., Markesbery, W. R., & Landfield, P. W. (2004). Incipient Alzheimer's disease: Microarray correlation analyses reveal major transcriptional and tumor suppressor responses. *Proceedings of the National Academy of Sciences of the United States of America*, *101*, 2173–2178.

Booij, B. B., Lindahl, T., Wetterberg, P., Skaane, N. V., Saebo, S., Feten, G., et al. (2011). A gene expression pattern in blood for the early detection of Alzheimer's disease. *Journal of Alzheimer's Disease*, *23*, 109–119.

Bouter, Y., Kacprowski, T., Weissmann, R., Dietrich, K., Borgers, H., Brauss, A., et al. (2014). Deciphering the molecular profile of plaques, memory decline and neuron loss in two mouse models for Alzheimer's disease by deep sequencing. *Frontiers in Aging Neuroscience*, *6*, 75.

Bradley-Whitman, M. A., & Lovell, M. A. (2013). Epigenetic changes in the progression of Alzheimer's disease. *Mechanisms of Ageing and Development*, *134*, 486–495.

Budovsky, A., Craig, T., Wang, J., Tacutu, R., Csordas, A., Lourenco, J., et al. (2013). LongevityMap: A database of human genetic variants associated with longevity. *Trends in Genetics*, *29*, 559–560.

Cadena-del-Castillo, C., Valdes-Quezada, C., Carmona-Aldana, F., Arias, C., Bermudez-Rattoni, F., & Recillas-Targa, F. (2014). Age-dependent increment of hydroxymethylation in the brain cortex in the triple-transgenic mouse model of Alzheimer's disease. *Journal of Alzheimer's Disease*, *41*, 845–854.

Callaway, E. (2012). Alzheimer's drugs take a new tack. *Nature, 489*, 13–14.

Carrasquillo, M. M., Crook, J. E., Pedraza, O., Thomas, C. S., Pankratz, V. S., Allen, M., et al. (2015). Late-onset Alzheimer's risk variants in memory decline, incident mild cognitive impairment, and Alzheimer's disease. *Neurobiology of Aging, 36*, 60–67.

Chen, K. D., Chang, P. T., Ping, Y. H., Lee, H. C., Yeh, C. W., & Wang, P. N. (2011). Gene expression profiling of peripheral blood leukocytes identifies and validates ABCB1 as a novel biomarker for Alzheimer's disease. *Neurobiology of Disease, 43*, 698–705.

Chouliaras, L., Mastroeni, D., Delvaux, E., Grover, A., Kenis, G., Hof, P. R., et al. (2013). Consistent decrease in global DNA methylation and hydroxymethylation in the hippocampus of Alzheimer's disease patients. *Neurobiology of Aging, 34*, 2091–2099.

Chung, S. J., Jung, Y., Hong, M., Kim, M. J., You, S., Kim, Y. J., et al. (2013). Alzheimer's disease and Parkinson's disease genome-wide association study top hits and risk of Parkinson's disease in Korean population. *Neurobiology of Aging, 34*, 2695.e1–2695.e7.

Condliffe, D., Wong, A., Troakes, C., Proitsi, P., Patel, Y., Chouliaras, L., et al. (2014). Cross-region reduction in 5-hydroxymethylcytosine in Alzheimer's disease brain. *Neurobiology of Aging, 35*, 1850–1854.

Coppieters, N., Dieriks, B. V., Lill, C., Faull, R. L., Curtis, M. A., & Dragunow, M. (2014). Global changes in DNA methylation and hydroxymethylation in Alzheimer's disease human brain. *Neurobiology of Aging, 35*, 1334–1344.

Cruchaga, C., Karch, C. M., Jin, S. C., Benitez, B. A., Cai, Y., Guerreiro, R., et al. (2014). Rare coding variants in the phospholipase D3 gene confer risk for Alzheimer's disease. *Nature, 505*, 550–554.

Cui, J. G., Li, Y. Y., Zhao, Y., Bhattacharjee, S., & Lukiw, W. J. (2010). Differential regulation of interleukin-1 receptor-associated kinase-1 (IRAK-1) and IRAK-2 by microRNA-146a and NF-kappaB in stressed human astroglial cells and in Alzheimer disease. *The Journal of Biological Chemistry, 285*, 38951–38960.

Davies, G., Harris, S. E., Reynolds, C. A., Payton, A., Knight, H. M., Liewald, D. C., et al. (2014). A genome-wide association study implicates the APOE locus in nonpathological cognitive ageing. *Molecular Psychiatry, 19*, 76–87.

Demuro, A., Parker, I., & Stutzmann, G. E. (2010). Calcium signaling and amyloid toxicity in Alzheimer disease. *The Journal of Biological Chemistry, 285*, 12463–12468.

Escott-Price, V., Bellenguez, C., Wang, L. S., Choi, S. H., Harold, D., Jones, L., et al. (2014). Gene-wide analysis detects two new susceptibility genes for Alzheimer's disease. *PLoS One, 9*, e94661.

Fehlbaum-Beurdeley, P., Jarrige-Le Prado, A. C., Pallares, D., Carriere, J., Guihal, C., Soucaille, C., et al. (2010). Toward an Alzheimer's disease diagnosis via high-resolution blood gene expression. *Alzheimer's & Dementia: The Journal of the Alzheimer's Association, 6*, 25–38.

Feng, X., Bai, Z., Wang, J., Xie, B., Sun, J., Han, G., et al. (2014). Robust gene dysregulation in Alzheimer's disease brains. *Journal of Alzheimer's Disease, 41*, 587–597.

Furney, S. J., Simmons, A., Breen, G., Pedroso, I., Lunnon, K., Proitsi, P., et al. (2011). Genome-wide association with MRI atrophy measures as a quantitative trait locus for Alzheimer's disease. *Molecular Psychiatry, 16*, 1130–1138.

Ghani, M., Pinto, D., Lee, J. H., Grinberg, Y., Sato, C., Moreno, D., et al. (2012). Genome-wide survey of large rare copy number variants in Alzheimer's disease among Caribbean hispanics. *G3 (Bethesda, Md.), 2*, 71–78.

Guerreiro, R., Bras, J., & Hardy, J. (2013). SnapShot: Genetics of Alzheimer's disease. *Cell, 155*, 968–968.e1.

Guerreiro, R. J., Lohmann, E., Kinsella, E., Brás, J. M., Luu, N., Gurunlian, N., et al. (2012). Exome sequencing reveals an unexpected genetic cause of disease: NOTCH3 mutation in a Turkish family with Alzheimer's disease. *Neurobiology of Aging, 33*, 1008.e17–1008.e23.

Guffanti, G., Torri, F., Rasmussen, J., Clark, A. P., Lakatos, A., Turner, J. A., et al. (2013). Increased CNV-region deletions in mild cognitive impairment (MCI) and Alzheimer's disease (AD) subjects in the ADNI sample. *Genomics, 102*, 112–122.

Han, G., Sun, J., Wang, J., Bai, Z., Song, F., & Lei, H. (2014). Genomics in neurological disorders. *Genomics, Proteomics & Bioinformatics, 12*, 156–163.

Han, G., Wang, J., Zeng, F., Feng, X., Yu, J., Cao, H. Y., et al. (2013). Characteristic transformation of blood transcriptome in Alzheimer's disease. *Journal of Alzheimer's Disease, 35*, 373–386.

Harris, S. E., Davies, G., Luciano, M., Payton, A., Fox, H. C., Haggarty, P., et al. (2013). Polygenic risk for Alzheimer's disease is not associated with cognitive ability or cognitive aging in non-demented older people. *Journal of Alzheimer's Disease, 39*, 565–574.

Hebert, S. S., Horre, K., Nicolai, L., Papadopoulou, A. S., Mandemakers, W., Silahtaroglu, A. N., et al. (2008). Loss of microRNA cluster miR-29a/b-1 in sporadic Alzheimer's disease correlates with increased BACE1/beta-secretase expression. *Proceedings of the National Academy of Sciences of the United States of America, 105*, 6415–6420.

Hebert, S. S., Wang, W. X., Zhu, Q., & Nelson, P. T. (2013). A study of small RNAs from cerebral neocortex of pathology-verified Alzheimer's disease, dementia with lewy bodies, hippocampal sclerosis, frontotemporal lobar dementia, and non-demented human controls. *Journal of Alzheimer's Disease, 35*, 335–348.

Heinzen, E. L., Need, A. C., Hayden, K. M., Chiba-Falek, O., Roses, A. D., Strittmatter, W. J., et al. (2010). Genome-wide scan of copy number variation in late-onset Alzheimer's disease. *Journal of Alzheimer's Disease, 19*, 69–77.

Holton, P., Ryten, M., Nalls, M., Trabzuni, D., Weale, M. E., Hernandez, D., et al. (2013). Initial assessment of the pathogenic mechanisms of the recently identified Alzheimer risk loci. *Annals of Human Genetics*, 77, 85–105.

Hooli, B. V., Kovacs-Vajna, Z. M., Mullin, K., Blumenthal, M. A., Mattheisen, M., Zhang, C., et al. (2014). Rare autosomal copy number variations in early-onset familial Alzheimer's disease. *Molecular Psychiatry, 19*, 676–681.

Israel, M. A., Yuan, S. H., Bardy, C., Reyna, S. M., Mu, Y., Herrera, C., et al. (2012). Probing sporadic and familial Alzheimer's disease using induced pluripotent stem cells. *Nature, 482*, 216–220.

Jahanshad, N., Rajagopalan, P., Hua, X., Hibar, D. P., Nir, T. M., Toga, A. W., et al. (2013). Genome-wide scan of healthy human connectome discovers SPON1 gene variant influencing dementia severity. *Proceedings of the National Academy of Sciences of the United States of America, 110*, 4768–4773.

Kamboh, M. I., Barmada, M. M., Demirci, F. Y., Minster, R. L., Carrasquillo, M. M., Pankratz, V. S., et al. (2012). Genome-wide association analysis of age-at-onset in Alzheimer's disease. *Molecular Psychiatry, 17*, 1340–1346.

Karch, C. M., Jeng, A. T., Nowotny, P., Cady, J., Cruchaga, C., & Goate, A. M. (2012). Expression of novel Alzheimer's disease risk genes in control and Alzheimer's disease brains. *PLoS One*, 7, e50976.

Kim, K. H., Moon, M., Yu, S. B., Mook-Jung, I., & Kim, J. I. (2012). RNA-Seq analysis of frontal cortex and cerebellum from 5XFAD mice at early stage of disease pathology. *Journal of Alzheimer's Disease, 29*, 793–808.

Kumar, P., Dezso, Z., MacKenzie, C., Oestreicher, J., Agoulnik, S., Byrne, M., et al. (2013). Circulating miRNA biomarkers for Alzheimer's disease. *PLoS One, 8*, e69807.

Lambert, J.-C., Ibrahim-Verbaas, C. A., Harold, D., Naj, A. C., Sims, R., Bellenguez, C., et al. (2013). Meta-analysis of 74,046 individuals identifies 11 new susceptibility loci for Alzheimer's disease. *Nature Genetics, 45*(12), 1452–1458.

Lashley, T., Gami, P., Valizadeh, N., Li, A., Revesz, T., & Balazs, R. (2014). Alterations in global DNA methylation and hydroxymethylation are not detected in Alzheimer's disease. *Neuropathology and Applied Neurobiology, 41*(4), 497–506.

Lau, P., Bossers, K., Janky, R., Salta, E., Frigerio, C. S., Barbash, S., et al. (2013). Alteration of the microRNA network during the progression of Alzheimer's disease. *EMBO Molecular Medicine, 5*, 1613–1634.

Lee, J. H., Kahn, A., Cheng, R., Reitz, C., Vardarajan, B., Lantigua, R., et al. (2014). Disease-related mutations among Caribbean Hispanics with familial dementia. *Molecular Genetics & Genomic Medicine, 2*, 430–437.

Lei, H. (2010). Amyloid and Alzheimer's disease. *Protein & Cell, 1*, 312–314.

Leidinger, P., Backes, C., Deutscher, S., Schmitt, K., Mueller, S. C., Frese, K., et al. (2013). A blood based 12-miRNA signature of Alzheimer disease patients. *Genome Biology, 14*, R78.

Lencz, T., Knowles, E., Davies, G., Guha, S., Liewald, D. C., Starr, J. M., et al. (2014). Molecular genetic evidence for overlap between general cognitive ability and risk for schizophrenia: A report from the Cognitive Genomics consorTium (COGENT). *Molecular Psychiatry, 19*, 168–174.

Li, G., Bekris, L. M., Leong, L., Steinbart, E. J., Shofer, J. B., Crane, P. K., et al. (2013). TOMM40 intron 6 poly-T length, age at onset, and neuropathology of AD in individuals with APOE epsilon3/epsilon3. *Alzheimer's & Dementia: The Journal of the Alzheimer's Association, 9*, 554–561.

Liang, W. S., Dunckley, T., Beach, T. G., Grover, A., Mastroeni, D., Ramsey, K., et al. (2008). Altered neuronal gene expression in brain regions differentially affected by Alzheimer's disease: A reference data set. *Physiological Genomics, 33*, 240–256.

Liang, D., Han, G., Feng, X., Sun, J., Duan, Y., & Lei, H. (2012). Concerted perturbation observed in a hub network in Alzheimer's disease. *PLoS One*, 7(7), e40498.

Lim, Y. Y., Villemagne, V. L., Laws, S. M., Pietrzak, R. H., Snyder, P. J., Ames, D., et al. (2014). APOE and BDNF polymorphisms moderate amyloid beta-related cognitive decline in preclinical Alzheimer's disease. *Molecular Psychiatry*. http://dx.doi.org/10.1038/mp.2014.123.

Liu, Q., Waltz, S., Woodruff, G., Ouyang, J., Israel, M. A., Herrera, C., et al. (2014). Effect of potent gamma-secretase modulator in human neurons derived from multiple presenilin 1-induced pluripotent stem cell mutant carriers. *JAMA Neurology, 71*, 1481–1489.

Logue, M. W., Schu, M., Vardarajan, B. N., Farrell, J., Lunetta, K. L., Jun, G., et al. (2014). Search for age-related macular degeneration risk variants in Alzheimer disease genes and pathways. *Neurobiology of Aging, 35*, 1510.e7–1510.e18.

Lu, T., Pan, Y., Kao, S. Y., Li, C., Kohane, I., Chan, J., et al. (2004). Gene regulation and DNA damage in the ageing human brain. *Nature, 429*, 883–891.

Lunnon, K., Ibrahim, Z., Proitsi, P., Lourdusamy, A., Newhouse, S., Sattlecker, M., et al. (2012). Mitochondrial dysfunction and immune activation are detectable in early Alzheimer's disease blood. *Journal of Alzheimer's Disease, 30*, 685–710.

Lunnon, K., & Mill, J. (2013). Epigenetic studies in Alzheimer's disease: Current findings, caveats, and considerations for future studies. *American Journal of Medical Genetics Part B, Neuropsychiatric Genetics, 162B*, 789–799.

Maes, O. C., Xu, S., Yu, B., Chertkow, H. M., Wang, E., & Schipper, H. M. (2007). Transcriptional profiling of Alzheimer blood mononuclear cells by microarray. *Neurobiology of Aging, 28*, 1795–1809.

Maloney, J. A., Bainbridge, T., Gustafson, A., Zhang, S., Kyauk, R., Steiner, P., et al. (2014). Molecular mechanisms of Alzheimer disease protection by the A673T allele of amyloid precursor protein. *The Journal of Biological Chemistry, 289*, 30990–31000.

Marchani, E. E., Bird, T. D., Steinbart, E. J., Rosenthal, E., Yu, C. E., Schellenberg, G. D., et al. (2010). Evidence for three loci modifying age-at-onset of Alzheimer's disease in early-onset PSEN2 families. *American Journal of Medical Genetics. Part B, Neuropsychiatric Genetics, 153B*, 1031–1041.

Mastroeni, D., Grover, A., Delvaux, E., Whiteside, C., Coleman, P. D., & Rogers, J. (2010). Epigenetic changes in Alzheimer's disease: Decrements in DNA methylation. *Neurobiology of Aging, 31*, 2025–2037.

Melville, S. A., Buros, J., Parrado, A. R., Vardarajan, B., Logue, M. W., Shen, L., et al. (2012). Multiple loci influencing hippocampal degeneration identified by genome scan. *Annals of Neurology, 72*, 65–75.

Moskvina, V., Harold, D., Russo, G., Vedernikov, A., Sharma, M., Saad, M., et al. (2013). Analysis of genome-wide association studies of Alzheimer disease and of Parkinson disease to determine if these 2 diseases share a common genetic risk. *JAMA Neurology, 70*, 1268–1276.

Nho, K., Corneveaux, J. J., Kim, S., Lin, H., Risacher, S. L., Shen, L., et al. (2013). Whole-exome sequencing and imaging genetics identify functional variants for rate of change in hippocampal volume in mild cognitive impairment. *Molecular Psychiatry, 18*, 781–787.

Proitsi, P., Lupton, M. K., Velayudhan, L., Hunter, G., Newhouse, S., Lin, K., et al. (2014). Alleles that increase risk for type 2 diabetes mellitus are not associated with increased risk for Alzheimer's disease. *Neurobiology of Aging, 35*, 2883.e3–2883.e10.

Ramanan, V. K., Risacher, S. L., Nho, K., Kim, S., Swaminathan, S., Shen, L., et al. (2014). APOE and BCHE as modulators of cerebral amyloid deposition: A florbetapir PET genome-wide association study. *Molecular Psychiatry, 19*, 351–357.

Rao, J. S., Keleshian, V. L., Klein, S., & Rapoport, S. I. (2012). Epigenetic modifications in frontal cortex from Alzheimer's disease and bipolar disorder patients. *Translational Psychiatry, 2*, e132.

Rodriguez-Ortiz, C. J., Baglietto-Vargas, D., Martinez-Coria, H., LaFerla, F. M., & Kitazawa, M. (2014). Upregulation of miR-181 decreases c-Fos and SIRT-1 in the hippocampus of 3xTg-AD mice. *Journal of Alzheimer's Disease, 42*, 1229–1238.

Rye, P. D., Booij, B. B., Grave, G., Lindahl, T., Kristiansen, L., Andersen, H. M., et al. (2011). A novel blood test for the early detection of Alzheimer's disease. *Journal of Alzheimer's Disease, 23*, 121–129.

Sanchez-Mut, J. V., Aso, E., Panayotis, N., Lott, I., Dierssen, M., Rabano, A., et al. (2013). DNA methylation map of mouse and human brain identifies target genes in Alzheimer's disease. *Brain, 136*, 3018–3027.

Schiepers, O. J., Harris, S. E., Gow, A. J., Pattie, A., Brett, C. E., Starr, J. M., et al. (2012). APOE E4 status predicts age-related cognitive decline in the ninth decade: Longitudinal follow-up of the Lothian Birth Cohort 1921. *Molecular Psychiatry, 17*, 315–324.

Sebastiani, P., Solovieff, N., Dewan, A. T., Walsh, K. M., Puca, A., Hartley, S. W., et al. (2012). Genetic signatures of exceptional longevity in humans. *PLoS One, 7*, e29848.

Sekar, S., McDonald, J., Cuyugan, L., Aldrich, J., Kurdoglu, A., Adkins, J., et al. (2015). Alzheimer's disease is associated with altered expression of genes involved in immune response and mitochondrial processes in astrocytes. *Neurobiology of Aging, 36*(2), 583–591.

Sheinerman, K. S., Tsivinsky, V. G., Abdullah, L., Crawford, F., & Umansky, S. R. (2013). Plasma microRNA biomarkers for detection of mild cognitive impairment: Biomarker validation study. *Aging, 5*, 925–938.

Sheinerman, K. S., Tsivinsky, V. G., Crawford, F., Mullan, M. J., Abdullah, L., & Umansky, S. R. (2012). Plasma microRNA biomarkers for detection of mild cognitive impairment. *Aging, 4*, 590–605.

Sherva, R., Tripodis, Y., Bennett, D. A., Chibnik, L. B., Crane, P. K., de Jager, P. L., et al. (2014). Genome-wide association study of the rate of cognitive decline in Alzheimer's disease. *Alzheimer's & Dementia: The Journal of the Alzheimer's Association, 10*, 45–52.

Shulman, J. M., Chen, K., Keenan, B. T., Chibnik, L. B., Fleisher, A., Thiyyagura, P., et al. (2013). Genetic susceptibility for Alzheimer disease neuritic plaque pathology. *JAMA Neurology, 70*, 1150–1157.

Sproul, A. A., Jacob, S., Pre, D., Kim, S. H., Nestor, M. W., Navarro-Sobrino, M., et al. (2014). Characterization and molecular profiling of PSEN1 familial Alzheimer's disease iPSC-derived neural progenitors. *PLoS One, 9*, e84547.

Stein, J. L., Medland, S. E., Vasquez, A. A., Hibar, D. P., Senstad, R. E., Winkler, A. M., et al. (2012). Identification of common variants associated with human hippocampal and intracranial volumes. *Nature Genetics, 44*, 552–561.

Sun, J., Feng, X., Liang, D., Duan, Y., & Lei, H. (2012). Down-regulation of energy metabolism in Alzheimer's disease is a protective response of neurons to the microenvironment. *Journal of Alzheimer's Disease, 28*, 389–402.

Sun, J., Song, F., Wang, J., Han, G., Bai, Z., Xie, B., et al. (2014). Hidden risk genes with high-order intragenic epistasis in Alzheimer's disease. *Journal of Alzheimer's Disease, 41*, 1039–1056.

Sun, X., Wu, Y., Gu, M., & Zhang, Y. (2014). miR-342-5p decreases ankyrin G levels in Alzheimer's disease transgenic mouse models. *Cell Reports, 6*, 264–270.

Szigeti, K., Kellermayer, B., Lentini, J. M., Trummer, B., Lal, D., Doody, R. S., et al. (2014). Ordered subset analysis of copy number variation association with age at onset of Alzheimer's disease. *Journal of Alzheimer's Disease, 41*, 1063–1071.

Szigeti, K., Lal, D., Li, Y., Doody, R. S., Wilhelmsen, K., Yan, L., et al. (2013). Genome-wide scan for copy number variation association with age at onset of Alzheimer's disease. *Journal of Alzheimer's Disease, 33*, 517–523.

Tan, L., Yu, J. T., Tan, M. S., Liu, Q. Y., Wang, H. F., Zhang, W., et al. (2014). Genome-wide serum microRNA expression profiling identifies serum biomarkers for Alzheimer's disease. *Journal of Alzheimer's Disease, 40*, 1017–1027.

Thambisetty, M., An, Y., & Tanaka, T. (2013). Alzheimer's disease risk genes and the age-at-onset phenotype. *Neurobiology of Aging, 34*, 2696.e1–2696.e5.

Tian, N., Cao, Z., & Zhang, Y. (2014). MiR-206 decreases brain-derived neurotrophic factor levels in a transgenic mouse model of Alzheimer's disease. *Neuroscience Bulletin, 30*, 191–197.

Verhaaren, B. F., Vernooij, M. W., Koudstaal, P. J., Uitterlinden, A. G., van Duijn, C. M., Hofman, A., et al. (2013). Alzheimer's disease genes and cognition in the nondemented general population. *Biological Psychiatry, 73*, 429–434.

Walker, M. P., LaFerla, F. M., Oddo, S. S., & Brewer, G. J. (2013). Reversible epigenetic histone modifications and Bdnf expression in neurons with aging and from a mouse model of Alzheimer's disease. *Age, 35*, 519–531.

Walter, S., Atzmon, G., Demerath, E. W., Garcia, M. E., Kaplan, R. C., Kumari, M., et al. (2011). A genome-wide association study of aging. *Neurobiology of Aging, 32*, 2109.e15–2109.e28.

Wang, J., Feng, X., Bai, Z., Jin, L. W., Duan, Y., & Lei, H. (2014). Chromosome 19p in Alzheimer's disease: When genome meets transcriptome. *Journal of Alzheimer's Disease, 38*, 245–250.

Wang, W. X., Rajeev, B. W., Stromberg, A. J., Ren, N., Tang, G., Huang, Q., et al. (2008). The expression of microRNA miR-107 decreases early in Alzheimer's disease and may accelerate disease progression through regulation of beta-site amyloid precursor protein-cleaving enzyme 1. *The Journal of Neuroscience, 28*, 1213–1223.

Webster, J. A., Gibbs, J. R., Clarke, J., Ray, M., Zhang, W., Holmans, P., et al. (2009). Genetic control of human brain transcript expression in Alzheimer disease. *American Journal of Human Genetics, 84*, 445–458.

Wong, H. K., Veremeyko, T., Patel, N., Lemere, C. A., Walsh, D. M., Esau, C., et al. (2013). De-repression of FOXO3a death axis by microRNA-132 and -212 causes neuronal apoptosis in Alzheimer's disease. *Human Molecular Genetics, 22*, 3077–3092.

Woodruff, G., Young, J. E., Martinez, F. J., Buen, F., Gore, A., Kinaga, J., et al. (2013). The presenilin-1 DeltaE9 mutation results in reduced gamma-secretase activity, but not total loss of PS1 function, in isogenic human stem cells. *Cell Reports, 5*, 974–985.

Zhang, B., Gaiteri, C., Bodea, L. G., Wang, Z., McElwee, J., Podtelezhnikov, A. A., et al. (2013). Integrated systems approach identifies genetic nodes and networks in late-onset Alzheimer's disease. *Cell, 153*, 707–720.

CHAPTER TWO

The Potential of Proteomics in Understanding Neurodegeneration

Ramavati Pal*, Jan Petter Larsen†, Simon Geir Moller*,†,1

*Department of Biological Sciences, St. John's University, New York, USA
†Norwegian Center for Movement Disorders, Stavanger University Hospital, Stavanger, Norway
[1]Corresponding author: e-mail address: mollers@stjohns.edu

Contents

1. Introduction 26
2. Cellular Mechanisms of Neurodegeneration 27
3. Fundamentals of Proteomics 28
   3.1 Common Proteomic Protocols 30
   3.2 Specifics of Proteomic Protocols 33
4. Application of Proteomics in Neurodegenerative Diseases 34
   4.1 Proteomic Approaches Using Animal Models 34
   4.2 Proteomic Approaches Using Human Cells and Tissue 42
   4.3 Proteomic Characterization of Biological Fluids for Biomarker Discovery 44
5. From Proteomics to Clinical Significance: Targeting Signaling Pathways in Neurodegenerative Diseases 47
6. Future Perspectives: Combinatorial Genomic, Metabolomics, and Proteomic Analysis 48

Acknowledgments 49

References 49

Abstract

Neurodegenerative diseases are a major health concern worldwide. Diseases such as Alzheimer's disease, Parkinson's disease, Huntington's disease, and amyotrophic lateral sclerosis, as well as many other diseases affecting the neuromuscular system, are a leading cause of disability in the aging population. Presymptomatic diagnosis of neurodegenerative disorders is challenging due to the lack of robust biomarkers. Likewise, the design of effective intervention strategies is limited because most neurodegenerative disorders are heterogeneous in nature. Reliable noninvasive biomarkers are therefore urgently needed to allow presymptomatic and accurate diagnosis, to track disease progression, to evaluate the effectiveness of new treatment regimens, and to ultimately design new therapeutic intervention strategies. Recent biological and technological advances within the field of proteomic promises to provide insight into global proteome changes in neurodegeneration, thus allowing increased understanding of

International Review of Neurobiology, Volume 121
ISSN 0074-7742
http://dx.doi.org/10.1016/bs.irn.2015.05.002

molecular pathways leading to neuronal cell death and the identification of biomarkers. The combination of gel-based techniques and mass spectrometry permits large-scale identification of peptide sequences in biological samples as well as the characterization of posttranslational protein modifications. The application of comparative high-throughput proteomic analyses in animal models and human tissues will aid in the identification of both diagnostic and prognostic biomarkers and will provide a platform for a future personalized medicine approach in neurodegeneration.

1. INTRODUCTION

Neurodegenerative diseases such as Alzheimer's disease (AD), Parkinson's disease (PD), Huntington's disease (HD), and amyotrophic lateral sclerosis (ALS) pose serious public health challenges that will increase in the next decades. The genetic causes and risk factors for several neurodegenerative diseases have been discovered (Pardo & van Duijn, 2005; Tsuang & Bird, 2002); however, most neurodegenerative diseases occur sporadically in an age-dependent manner (David, 2006). Neurodegenerative disorders are generally characterized by the gradual loss of motor, sensory, and perceptual functions and associated cognitive and behavioral deficits, attributed to neuronal cell death (Coppede, Mancuso, Siciliano, Migliore, & Murri, 2006). Differences in origin and the role of both genetic and environmental factors in onset and progression of neurodegenerative diseases complicate our understanding of the pathogenic mechanisms involved (Migliore & Coppede, 2002).

A pathological hallmark of many neurodegenerative diseases is the accumulation and aggregation of misfolded proteins resulting in intracellular inclusions within neurons (Taylor, Hardy, & Fischbeck, 2002). Although the aggregation mechanisms and the associated biochemical pathways are poorly understood, numerous gene mutations, that segregate with neurodegenerative disease cohorts, have been identified and the encoded proteins are therefore assumed to be associated with neurodegeneration (Fig. 1). Neuronal proteomic approaches, such as comparative proteome analysis and protein posttranslational modification (PTM)/network characterization, will undoubtedly increase our knowledge of how these proteins influence neuronal cell death. By combining these proteomic approaches with well-defined clinical features, we will not only further our understanding of disease onset and progression but also identify diagnostic and prognostic biomarkers invaluable in the development of novel therapeutic intervention

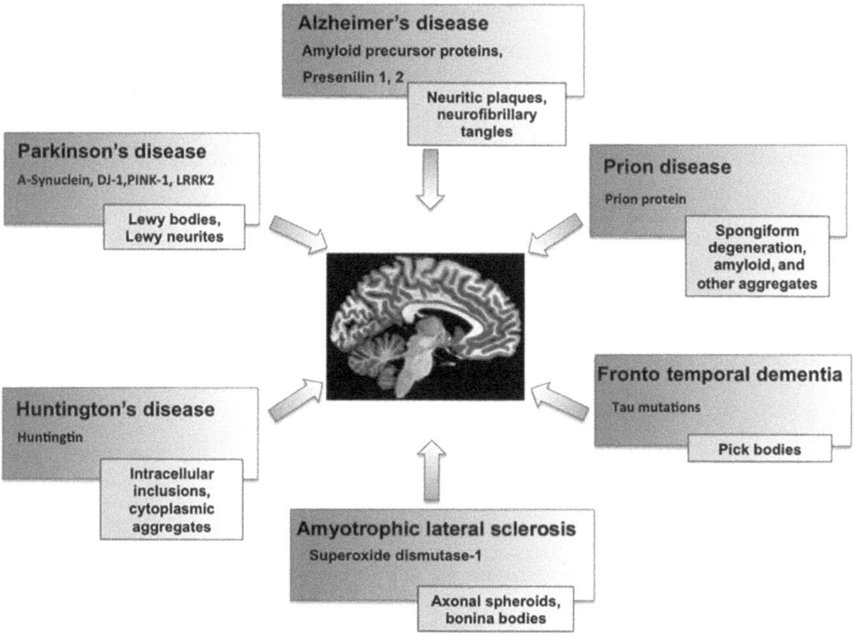

Figure 1 Various neurodegenerative diseases with their main characteristic pathologies.

strategies. In this chapter, we provide a comprehensive review of the diverse proteomic approaches that have been used to dissect some of the underlying pathogenic mechanisms in neurodegeneration and the impact that these findings may have in the field.

2. CELLULAR MECHANISMS OF NEURODEGENERATION

The intracellular and extracellular aggregation of mutated and/or misfolded proteins is a common feature of many neurodegenerative diseases. In AD, neuronal cell death is associated with the deposition of extracellular amyloid plaques, consisting of amyloid β peptides (Aβ, ($A\beta_{40}$ and $A\beta_{42}$)) and intracellular neurofibrillary tangles (Kawamata et al., 1998; Zilkova, Koson, & Zilka, 2006). Although the exact mechanism leading to AD is not fully understood, various studies in AD models have started to unravel some of the pathogenic processes underlying the disease (Hardy & Selkoe, 2002; Maccioni, Farias, Morales, & Navarrete, 2010; Mudher & Lovestone, 2002).

The main characteristic of PD is the loss of dopaminergic neurons in the substantia nigra and formation of Lewy bodies enriched in the protein α-synuclein (SNCA) (Martin et al., 2012). Although the majority of PD cases are sporadic and age-related, mutations in several genes have been identified in early-onset familial forms of PD including SNCA, Parkin (PARK2), ubiquitin carboxyl-terminal esterase L1 (UCHL1), Parkinson protein 7 (PARK7, DJ-1), PTEN-induced putative kinase 1 (PINK1), leucine-rich repeat kinase 2 (LRRK2), SNCA-interacting protein (SNCAIP), and β-glucocerebrosidase (GBA) (Pankratz & Foroud, 2007).

The underlying characteristics of HD, a trinucleotide repeat disorder, are slightly different from that of AD and PD and include polyglutamine expansions within the Huntingtin (HTT) protein. Mutant HTT (mHTT) causes cellular dysfunction and neurodegeneration, probably through a combination of toxic gain- and loss-of-function mechanisms (Hands & Wyttenbach, 2010).

As for other neurodegenerative disorders, ALS is characterized by neuronal cell death preceded by cellular shrinkage and axonal swelling (Tandan & Bradley, 1985a, 1985b). Less than 10% of ALS cases are familial having mutations in the gene superoxide dismutase 1 (SOD1) involved in the regulation of free radical formation (Battistini et al., 2005).

Although the proteins affected in AD, PD, HD, and ALS are different, there are several common mechanisms, including the formation of oligomeric protein aggregates and increased oxidative stress, that ultimately lead to neuronal cell death. Studying the cellular and molecular pathways leading to aggregation and neurotoxicity promises to aid in the identification of both general and specific targets for therapeutic intervention strategies in a variety of neurodegenerative diseases.

3. FUNDAMENTALS OF PROTEOMICS

The Australian scientist Marc Wilkins coined the term "Proteomics" to represent all proteins expressed by a genome that defines an organism (Wilkins et al., 1996). Proteomics is now widely known as the study of protein expression, function, and interaction in healthy, as well as in diseased biological systems (Anderson & Anderson, 1998). A fundamental difference between proteomics and genomics is that while the genome is relatively stable, the proteome of a given genome is constantly in flux depending on developmental and environmental cues (Tyers & Mann, 2003). Hence, proteomic technologies can dissect systematic and global proteome changes within a biological system, in response to intrinsic or extrinsic perturbations such as those that occur in neurodegeneration (Fiehn, 2001).

Depending on the application, proteomics can be divided into several categories: (i) *Functional proteomics* focuses on producing information related to protein expression levels, PTMs, and protein activity. It can also be combined with systematic biochemical network analysis, using mathematical modeling or other system biology tools. Coupling mass spectrometry to protein microarray technology can also aid in assessing protein–protein interaction profiles (Suchkov, Gnatenko, Kostiushev, Krynskii, & Pal'tsev, 2013). *Kinomics*, which is a subbranch of functional proteomics, focuses on the identification of kinase targets and the elucidation of the cellular pathways and networks in which they reside (Kolch & Pitt, 2010; Manning, Whyte, Martinez, Hunter, & Sudarsanam, 2002). (iii) *Chemical proteomics* represents the integration of biochemistry, cell biology, organic synthesis, and MS (Savino, Paduano, Preiano, & Terracciano, 2012). A typical experiment involves three main steps: (a) immobilization of a molecule, often a chemically modified drug, to a solid-state support; (b) incubation of the molecule/solid state with a tissue extract/cell lysate to allow protein binding; and (c) elution of interacting proteins followed by MS analysis. Alternatively, tissue or cell lysates can be preincubated with the target molecule prior to introduction of the solid-state support (Sin, Meng, Auth, & Crews, 1998). This approach offers the opportunity to screen for multiple potential biomarkers and drug targets. (iv) *Clinical proteomics*, or *translational proteomics*, is defined as all MS-based preclinical, clinical, and molecular medicine approaches aimed at discovering and understanding the role of proteins in pathology. Ultimately, the aim of clinical proteomics is to help facilitate early disease diagnosis, prediction of disease progression, the identification of new therapeutic targets, and the evaluation of treatment responses (Beretta, 2007; Matt, Fu, Fu, & Van Eyk, 2008). (v) *Redox proteomics* focuses on the identification and quantification of redox-based proteome changes leading to a better understanding of protein oxidative modifications such as carbonylation, tyrosine nitration, and glutathionylation under pathological conditions (Butterfield & Dalle-Donne, 2012; Ghezzi & Bonetto, 2003; Rinalducci, Murgiano, & Zolla, 2008). (vi) *Diglycine (DiGly) capture proteomics* studies protein degradation facilitated by the ubiquitin–proteasome system (UPS) (Kim et al., 2011). In the UPS pathway, ubiquitin is activated and transferred to substrates via an E1–E2–E3 cascade (Ye & Rape, 2009). The isopeptide bond between the ubiquitin C-terminal glycine and the lysine ε-amino group in UPS substrates allows detection of ubiquitylated targets by MS, because trypsin treatment of ubiquitin conjugates yields a characteristic "diGly trace" due to cleavage of the ubiquitin C-terminal Arg-Gly-Gly sequence (Peng et al., 2003).

3.1 Common Proteomic Protocols

The various proteomic approaches share common features; however, depending on the application and biological question, the approach is often tailored to suit specific needs. The different steps and methodologies of proteomic experiments are depicted in Fig. 2.

Sample preparation is the main source of variation in the results of proteomic experiments; hence, consistent protocols should be followed. The first step in all proteomic experiments involves mechanical disruption and/or enzymatic digestion of the biological material. After tissue disruption, cells are usually lysed with detergents to extract the proteins (Yu & Cohen, 2004). For more targeted proteome studies, subproteomes or a subset of proteins can be isolated from total cell extracts before analysis (Yu & Cohen, 2004). Methodologies for this purpose include enrichment of hydrophobic or membrane-bound proteins using gradient detergent extractions, affinity-based purification, immunoprecipitation, and laser capture microdissection. In terms of biological fluids, sample preparation often requires removal of interfering components and/or fractionation prior to MS analysis.

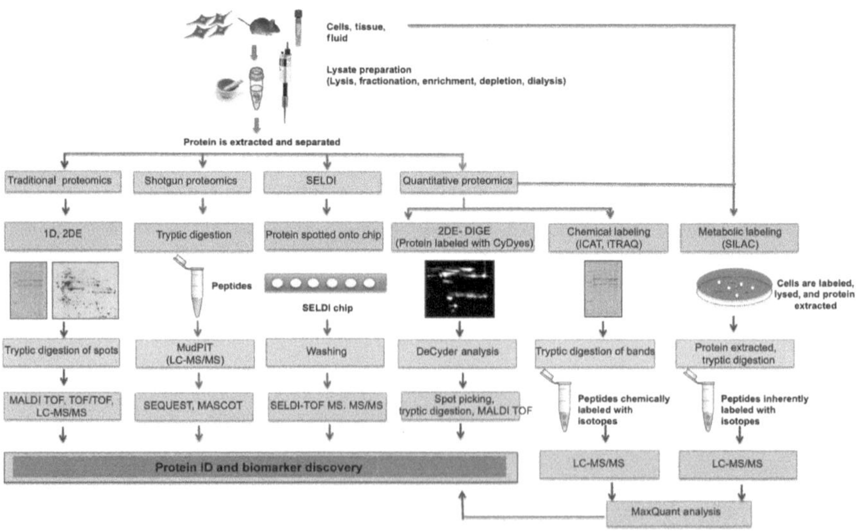

Figure 2 Schematic representation of the different steps involved in specific proteomic approaches.

Protein separation is required to reduce the complexity of subproteomes. Traditionally, proteins are separated by gel-based techniques such as two-dimensional gel electrophoresis (2-DE), in combination with MS (Hanash, 2001). In 2-DE, proteins are resolved according to their p*I* in the first dimension and their molecular weights in the second dimension (SDS-PAGE) (Gorg, Weiss, & Dunn, 2004). Protein spots can then be detected on the gel by a variety of protein staining methods followed by excision from the gel and digestion into peptides, typically using trypsin. For direct comparison of two protein samples, Difference In Gel Electrophoresis (DIGE) is used, which relies on the protein samples being labeled with different fluorescent dyes (Ünlü, Morgan, Minden, 1997). The differentially labeled protein samples are mixed and subjected to separation in the same 2-DE gel allowing reliable quantification of differences in protein abundance between different samples. Another gel-based proteomic approach uses isotope affinity tags to increase the sensitivity of detection of smaller peptide fragments. In this method, isotope-coded affinity tags (ICAT) selectively label cysteine residues of peptide fragments following tryptic digest of the protein sample (Gygi et al., 1999). The labeled fraction is then selectively analyzed. A recent development, the iTRAQ technique (Applied Biosystems, Foster City, CA), enables every tryptic fragment to be labeled, hence increasing the confidence in protein identification (Aggarwal, Choe, & Lee, 2006; Zieske, 2006).

In addition to electrophoretic techniques, various gel-free approaches exist that rely on liquid chromatography (LC) separation procedures with automated tandem MS, enabling large-scale "shotgun" sequencing of complex mixtures (Kislinger & Emili, 2003). Multidimensional protein identification technology (MudPIT), an important extension of LC (Washburn, Wolters, & Yates, 2001), has proven to be effective for investigating global changes in protein expression as a function of development and disease (Koller et al., 2002; Le Roch et al., 2004; Schirmer, Florens, Guan, Yates, & Gerace, 2003). MudPIT relies on the combination of different stationary phases for optimal resolution in the separation of highly complex protein mixtures (Henkel et al., 2012).

Protein identification can be achieved by coupling chromatographic technologies on-line with MS. Proteins or protein mixtures are first digested with an endopeptidase, usually trypsin, and the masses of the resulting peptides are accurately measured by MS. MS is a robust technique for detecting the mass of charged protein species. There are many excellent reviews devoted to MS and we will only describe the basics in this chapter

(Bantscheff, Lemeer, Savitski, & Kuster, 2012; Guerrera & Kleiner, 2005; Wu & MacCoss, 2002; Yates, Ruse, & Nakorchevsky, 2009). The first step in MS is to ionize and vaporize the analytes which can be achieved by two means: matrix-assisted laser desorption ionization (MALDI) and electrospray ionization (ESI) (Ho et al., 2003). The peptide or protein ions are then introduced into the mass analyzer. There are various types of mass analyzers such as ion trap, triple quadrupole, time of flight (TOF), and Fourier transform ion cyclotron (FTICR) that act differently from each other in terms of ion separation, mass accuracy, and resolution. Tandem MS (MS-MS) analyzes the daughter or fragment ions of the parent ions whose mass was determined in first MS dimension. In these collision-induced dissociation tandem MS approaches, the applied energy favors fragmentation of the peptide (amide) bond over amino acid side chains in the parent ion, thereby producing daughter ions, from which the amino acid sequence can be derived. Surface-enhanced laser desorption/ionization (SELDI-MS), derived from MALDI-MS, combines retention chromatography in solid phase (ion exchange, RP, affinity surfaces) with TOF-MS detection of proteins and is normally employed for the analysis of protein patterns with diagnostic value (Issaq, Veenstra, Conrads, & Felschow, 2002).

Once the mass spectral data have been obtained, they need to be analyzed using bioinformatic tools. The development of search engines, such as SEQUEST and MASCOT, correlate actual tandem mass spectra with predicted tandem mass spectra based on species-specific amino acid sequences from protein or genomic databases (Eng, McCormack, & Yates, 1994; Fenyo & Beavis, 2003). The fact that these different programs do not work in exactly the same way can result in different peptide matches and coverage. Because of this, results from MS-based approaches most often require further validation by either Western blot analysis or ELISA.

Protein quantification can be performed using proteomic techniques such as isotopic (ICAT), isobaric (Tandem Mass Tag), or iTRAQ labeling of peptides (Ross et al., 2004; Thompson et al., 2003). Among the isotopic labeling approaches, stable isotope labeling with amino acids in cell culture (SILAC) uses MS to detect differences in protein abundance among samples using nonradioactive isotopes (Ong et al., 2002). Label-free quantitative data can also be obtained using the multiple reaction monitoring (MRM) scan mode in a triple quadrupole MS system, which allows high levels of sensitivity and quantification of molecules of interest (Lehnert et al., 2012). Non-MS-based protein identification techniques, such as antigen microarray technologies, enable the large-scale profiling of the specificity of antibody responses against autoantigens, tumor antigens, and microbial antigens. Antibody profiling provides great insight into pathogenesis and will enable the development of diagnostic tests in the clinic (Joos et al., 2000).

3.2 Specifics of Proteomic Protocols

Proteomic technologies can be modified or combined to make the processing faster and less labor intensive. For example, in a typical LC-MS/MS experiment, the subproteome is cleaved into peptides using tryptic digest procedures and these peptides are separated by LC and then ionized by ESI and subsequently subjected to tandem MS. Often, the separated peptides are mixed with a MALDI matrix and spotted onto a MALDI plate which is then subjected to tandem MS. The benefit of this approach is that the entire analysis is much more automated compared to 2-DE. However, obtaining quantitative data from these experiments is challenging because the height/size of the peptide ion peaks after analysis in LC-MS/MS does not necessarily correlate with the amount/quantity of the protein in the subproteome due to differential ionization efficiency and variable detector sensitivity. Research has addressed this disadvantage in high-throughput proteomics and developed a novel method, "the stable isotype dilution assay" (Al-Taher et al., 2013; Varga et al., 2012). This method includes the introduction of a stable isotope label, such as deuterium 13C, into the sample, which is then subjected to MS analysis. The isotopically labeled peptide serves as an accurate mass standard and forms the basis for the determination of relative or absolute abundance. Examples of this approach include ICAT and iTRAQ, as mentioned previously. It is now possible to use eight separate probes to isotopically label ε-amino lysyl groups. Within this approach, advanced analytical software is required to link isotopic data from multiple peptides to estimations of protein concentrations. The development of novel methods for quantification, without the use of isotope labeling, is currently being developed (Haqqani, Kelly, & Stanimirovic, 2008).

Another approach, SELDI-MS, is mainly used in studies involved in the identification of biomarkers. In SELDI-MS, proteins or peptides can be concentrated through biochemically active materials, immobilized on a variety of plates or chips. The protein binding is mainly based on chelation, charge, and hydrophobicity. Moreover, organic molecules can also be used to coat the SELDI plate in order to capture protein receptors, antibodies, or sometimes nucleic acids. After incubation, the washing of the plates removes weakly bound proteins and thus mimics affinity chromatography. Although SELDI represents a relatively fast and easy method, only a single MS dimension is used and therefore it's use is limited (Albrethsen, 2007).

Another technique of particular interest to pathologists is tissue-based MS that can profile proteins *in situ* from different tissue regions tissue (Cornett, Reyzer, Chaurand, & Caprioli, 2007). The methods previously described, however, are those that have been more frequently used so far to study human neurodegenerative diseases.

4. APPLICATION OF PROTEOMICS IN NEURODEGENERATIVE DISEASES

Different proteomic approaches have identified more than 300 unique proteins associated with neurodegeneration and psychiatric disorders. We have summarized the recent findings from these approaches in Table 1. These proteins are proposed to be involved in metabolism, cytoskeleton formation, signal transduction, transport, and detoxification (Fountoulakis & Kossida, 2006). In this section, we describe the use of proteomic approaches in neurodegenerative disease research from different sources such as animal models, cells, tissue, and biological fluids.

4.1 Proteomic Approaches Using Animal Models

Proteomic findings from PD animal models have recently provided insights into numerous pathways that may be related to disease pathogenesis. Parkin knock-out (KO) and A30P SNCA transgenic mice models have been developed to mimic human familial PD. A total of 87 differentially expressed proteins in parkin KO mice, relative to controls, have been identified using 2DE-DIGE (Periquet, Corti, Jacquier, & Brice, 2005). Indeed, the oxidative modification of proteins has been shown to lead to loss of function (Aksenov, Aksenova, Carney, & Butterfield, 1997; Butterfield, 2004; Hensley et al., 1995; Lauderback et al., 2001). For example, redox proteomic approach has been applied to the brains of A30P SNCA transgenic mice. The proteins lactate dehydrogenase 2, carbonic anhydrase 2, and α-enolase had significantly higher levels of carbonylation and very low activity in A30P mice relative to controls, highlighting the importance of oxidative stress in PD (Periquet et al., 2005). Recently, an unbiased proteomic screen was carried out to identify synaptic protein changes in αβγ-synuclein KO mice brains to gain further insight into the molecular functions of synuclein (Westphal & Chandra, 2013). This study showed that synuclein can induce

Table 1 Proteomic Studies on Various Sample Sources in Neurodegenerative Disease Research

Sample Source	Disease	Proteomic Methodologies	Main Findings	References
Animal models				
1-Methyl-4-phenyl-1,2,3,6-tetrahydropyridine (MPTP)-treated monkeys	PD	2DE, MALDI-TOF, LC-MS/MS	Energy metabolism, protective mechanisms, and signal transduction are altered in response to MPTP	Campello et al. (2013)
6-OHDA-parkinsonian rat model	PD	2DE, MALDI-TOF-MS	Proteome alterations indicate remodeling processes of dendrites, axons, and synapses in PD pathogenesis	Lessner et al. (2010)
MPTP-treated mice	PD	LC-FTICR-MS	Use of transcriptomics and proteomics to study mitochondrial dysfunction	Smith (2009)
6-Hydroxydopamine-treated mice model	PD	LC-MS/MS	Study of neuropeptides in mediating the central nervous system failure associated with PD	Nilsson et al. (2009)
A53T alpha-synuclein Drosophila model	PD	LC-MS/MS, LTQ-FT-MS, SCX chromatography	Role of actin cytoskeleton and mitochondria in the development of late symptoms in PD	Xun, Sowell, Kaufman, and Clemmer (2008)
Transgenic mice brain overexpressing mutated human amyloid precursor protein	AD	Biotin-switch affinity/ nano-LC-MS/MS	Involvement of S-nitrosylation in the pathogenesis of Alzheimer's disease	Zareba-Koziol, Szwajda, Dadlez, Wyslouch-Cieszynska, and Lalowski (2014)

Continued

Table 1 Proteomic Studies on Various Sample Sources in Neurodegenerative Disease Research—cont'd

Sample Source	Disease	Proteomic Methodologies	Main Findings	References
Triple transgenic (PS1M146V/APPSwe/TauP301L) mouse model	AD	2DE-DIGE	Clusterin and α-2-macroglobulin could be potential biomarker for early AD diagnosis	Sui et al. (2014)
Bleomycin hydrolase −/− and +/+ mouse model	AD	2DE, MALDI-TOF/TOF	Role of hyperhomocysteinemia and bleomycin hydrolase in neurodegeneration	Suszynska-Zajczyk, Luczak, Marczak, and Jakubowski (2014)
Transgenic amyloid precursor protein/presenilin-1 (Aβpp/ps-1) AD mouse model	AD	LC-MS/MS	Study of splenocyte proteome map to gain insights into mechanisms of peripheral immunity in AD.	Cao and Robinson (2014)
Transgenic TBA42 mice expressing AβpE3-42	AD	LC-MS/MS	AD-typical pathways detectable prior to behavioral changes	Yang, Wittnam, Zubarev, and Bayer (2013)
Transgenic mouse model expressing NSE-human wild-type tau	AD	2DE, ESI-Q-TOF-MS/MS	Proteomic study of mechanisms associated with neuronal degeneration in AD	Chang et al. (2013)
Mouse model expressing mutant HTT	HD	LC-MS/MS	Study of Huntington proteomic interactome in mammalian brain	Shirasaki et al. (2012)
Wild-type and mutant *HTT* knock-in mice	HD	LC-MS/MS	Overexpression inhibits cap-dependent translation of a reporter mRNA in an *in vitro* system	Culver et al. (2012)

HD knock-in mouse models (HdhQ150 and HdhQ92)	HD	SELDI-TOF	Proteomic changes reflect HD pathology and implicate mitochondrial changes	Deschepper, Hoogendoorn, Brooks, Dunnett, and Jones (2012)
HD (R6/2 transgenic) mice model	HD	2DE, MALDI-TOF	Study of protein expression and oxidative modifications in HD brain	Perluigi, Swomley, and Butterfield (2014)
hSOD1G93A mouse model	ALS	MALDI imaging mass spectrometry	Pathological alterations in brain regions (facial/trigeminal nuclei)	Acquadro et al. (2014)
Transgenic mice carrying wild-type human SOD1	ALS	LC-MS/MS	Galectin-3 may be a potential biomarker for ALS	Zhou et al. (2010)
Wobbler mouse (genotype wr/wr)	ALS	2DE-DIGE, LC-MS/MS	Study of muscle proteome affected by motor neuron degeneration	Staunton, Zweyer, Swandulla, and Ohlendieck (2012)
Mouse model, G127X	ALS	2DE-DIGE, LC-MS/MS	Study of spinal cord proteome in ALS	Bergemalm et al. (2009)
Human cells and tissues				
Neuroblastoma cells SH-SY5Y	PD	LC-MS/MS	Study of interaction between dopamine and α-synuclein	Takano et al. (2013)
Neuroblastoma cells SH-SY5Y	PD	SILAC, LC-MS/MS	Study of neuronal mitochondria using quantitative proteomics	Villeneuve, Tiede, Morsey, and Fox (2013)
Human postmortem brain	PD	LC-MS/MS	Mass spectrometry-based peptidomics to understand the brain and PD	Cafe-Mendes, Ferro, Britto, and Martins-de-Souza (2014)

Continued

Table 1 Proteomic Studies on Various Sample Sources in Neurodegenerative Disease Research—cont'd

Sample Source	Disease	Proteomic Methodologies	Main Findings	References
Human postmortem brain	PD	Homogenizer, low-speed centrifugation	Enrichment of neuromelanin granules and synaptosomes from substantia nigra and downstream proteomic applications	Plum et al. (2013)
Human postmortem brain cortical tissue	PD	LC-MS/MS	Identification of proteins in Lewy body inclusions	Xia et al. (2008)
T-lymphocytes	PD	2DE, LC-MS/MS	Dopaminergic stimulation has important effects on T-cell proteome under long-term treatment	Alberio et al. (2012)
Human postmortem brain	AD	iTRAQ, LC-MS/MS, Western blot	Quantitative phosphoproteomic analysis of neurofilaments in AD	Chiang et al. (2007)
Human postmortem brain	AD	iTRAQ, LC-MS/MS	Abeta and human amylin exert toxicity via mitochondrial dysfunction	Lim et al. (2010)
Neuroblastoma cells SH-SY5Y	AD	2DE-DIGE, MALDI-TOF/TOF, LC-MS/MS	Heightened/prolonged IL-18 levels can drive protein changes in AD pathogenesis	Sutinen, Pirttila, Anderson, Salminen, and Ojala (2012)
Substantia nigra and cortex region of human postmortem brain	AD	2-DE, Pro-Q Diamond, ESI quadrupole TOF LC-MS/MS	Phosphoproteome profiling	Zahid, Oellerich, Asif, and Ahmed (2012)

Hippocampal tissue of human postmortem brain	AD	2-DE, Orbitrap MS/MS	Quantitative analysis of phosphorylated proteins	Di Domenico et al. (2011)
Postmortem brain	HD	2DE, MALDI-TOF, Western blot	Role of oxidative stress to specific macromolecules in HD progression	Riederer, Schiffrin, Kovari, Bouras, and Riederer (2009)
HEK 293 cells	HD	Affinity and size exclusion chromatography, MS	Proteomics to reveal that cleavage of Huntingtin at residue Arg(167) mediates mutant Huntingtin toxicity in HD	Alberio et al. (2010)
HEK 293FT cells, striatal neuronal progenitor cells	HD	iTRAQ, HPLC cation exchange chromatography, LC-MS/MS	Key cellular functions and networks may be disrupted by abnormal interactions of expanded huntingtin due to polyglutamine repeats	Zabel and Klose (2004)
HEK 293 cells	HD	2D-nano-LC-MS/MS	Association of phosphorylation sites and mutant Huntingtin toxicity	Mitsui, Doi, and Nukina (2006)
Biological fluids				
CSF	PD	iTRAQ, LC-MS/MS, MRM	Identification of biomarkers in CSF from PD patients	Lehnert et al. (2012)
Serum	PD	2D LC-MS/MS, iTRAQ	Quantitative analysis of serum proteins in PD	Wijte et al. (2012)

Continued

Table 1 Proteomic Studies on Various Sample Sources in Neurodegenerative Disease Research—cont'd

Sample Source	Disease	Proteomic Methodologies	Main Findings	References
Serum	PD	MALDI-TOF-MS, magnetic bead cation exchange	Identification of serum protein biomarkers in PD	Li et al. (2011)
Plasma	PD	2-DE, MALDI-TOF	Revealed amyloid P component as a plasma biomarker in PD	Ringman et al. (2012)
CSF	PD	SELDI-TOF-MS	Proteomic profiling of CSF in PD	Dalrymple et al. (2007)
Serum	PD	2-DE, MALDI-TOF	Serum proteomic analysis of Chinese PD patients	Saez-Valero, Barquero, Marcos, McLean and Small (2000)
CSF	PD	LC-MALDI-TOF/TOF	Targeted quantitative proteomic analysis of CSF in PD and AD	Jahn et al. (2011)
CSF	AD	SELDI-TOF-MS	Study of level of amyloid-beta 1–42 in CSF in AD	Albertini et al. (2012)
Plasma	AD	2-DE, MALDI, Western blot	Amyloid-beta 1–42—a putative biomarker for AD	Cocciolo et al. (2012)
CSF	AD	MALDI-TOF/MS, LC-MS/MS	Biomarkers of AD in CSF	Turk (2006)
Plasma	AD	Anion exchange, reversed phase chromatography, 2-DE, MS	Application of novel multidimensional plasma protein separation technique to identify biomarkers	Boysen and Hearn (2001)

CSF	AD	LC-ESI-MS-TOF	CSF proteome change in familial AD	Bazan and Lukiw (2002)
Plasma	AD	2-DE, MALDI-TOF	Plasma biomarker in brain atrophy	Berciano, Infante, Mateo, and Combarros (2002)
CSF	AD	Capillary electrophoresis-MS (CE-MS)	CSF biomarker evaluation for AD	Yuan et al. (1995)
Plasma	AD	2-DE, MALDI	Plasma proteomics to identify biomarkers for AD	Wijesekera and Leigh (2009)
Serum	AD	LC-MS/MS, ELISA	Assessment of serum biomarkers for AD	Zahid et al. (2012)
CSF	HD	Electrospray ionization (ESI)	Proteomics for characterizing lipid-metabolizing proteins in human CSF	Zhang et al. (2010)
Plasma, CSF	HD	2-DE, MALDI-TOF, LC-MS/MS, ELISA	Multiplatform proteome profiling revealing plasma and CSF changes with HD progression	Bonifati, Oostra, and Heutink (2004)
Serum	HD	MALDI-TOF-MS	Increased prothrombin, apolipoprotein A-IV, and haptoglobin in CSF of HD patients	Huang et al. (2011)

membrane curvature, suggesting a new role for this central PD protein. Zhang and colleagues used 1-methyl-4-phenyl-1,2,3,6-tetrahydropyridine (MPTP)-induced PD mouse models to study comparative proteome changes in several brain regions including striatum, cerebellum, and the cortex, employing two-dimensional LC-MS/MS (Zhang et al., 2010). From this study, it was discovered that the ubiquitin-specific protease (USP9X), a deubiquination enzyme involved in the protection of proteins from degradation, and the promotion of the TGF-β pathway are most likely involved in neurodegeneration (Vijitruth et al., 2006; Zhang et al., 2010). 2DE-DIGE has also been employed to study the mitochondrial proteome in AD. Triple transgenic (3xTg)-AD mice, overexpressing three AD mutations (APPswe, PS1M146V, and TauP301L), were analyzed before senile plaque development and 23 differentially expressed mitochondrial proteins were identified thus giving insight into AD progression (Chou et al., 2011). The hippocampus of another AD mouse model (5XFAD), containing five early-onset familial AD mutations, was analyzed using quantitative proteomic approach revealing proteins involved in early stages of AD pathology such as ApoE, ApoJ, and nicastrin (Hong et al., 2013). Recently, the murine HD model R6/2 was studied at various ages and stages of disease (early-, intermediate-, and late-stage HD) using a large 2-DE and MS-based proteomic approach (Zabel et al., 2009). Findings from this study include the upregulation of proteins involved in glycolysis/gluconeogenesis at 2 weeks of age preceding the onset of HD, while the actin cytoskeleton was downregulated. By contrast, at later HD stages (12 weeks of age), proteasomal proteins were upregulated. Combined, this suggests large variations in protein expression during the course of HD (Zabel et al., 2009).

Although proteomic research, using a variety of neurodegenerative disease animal models, has yielded interesting data, it may be challenging to draw direct parallels between animal and human proteome dynamics. Because of this combinatorial animal and human proteomic approaches should be employed.

4.2 Proteomic Approaches Using Human Cells and Tissue

Proteomic approaches have had limited success in the study of the whole brain and larger brain regions. The mixed populations of distinct cell types in brain make it challenging to understand the cellular origin of a particular protein change. To reduce the complexity of the brain proteome, many subcellular fractionation methods have been developed. For examples, neurotransmitter receptors are enriched in postsynaptic fractions (Li et al., 2008), neurofilaments can be enriched from the frontal cortex from frozen postmortem human AD and control brains (Rudrabhatla, Grant, Jaffe, Strong, & Pant, 2010), and LBs can be enriched from postmortem brains from patients with an LB variant of AD (Xia et al., 2008). A number of proteomic studies have used both 2-DE and 2DE-DIGE to examine differential protein characteristics in human postmortem tissue from pathological and nonpathological subjects. Domenico and colleagues used hippocampal tissue from control and AD subjects to identify phosphoproteins by employing 2-DE with the phosphoprotein-specific stain Pro-Q diamond and from this identified 17 altered phosphoproteins (Di Domenico et al., 2011). Comparative 2-DE analysis on the substantia nigra from PD and control cohorts has not only revealed protein changes related to oxidative stress and mitochondrial dysfunction but also increased levels of cytosolic dipeptidase 2 (CNDP2) in PD subjects (Licker et al., 2012). Interestingly, CNDP2 has to date not been reported to be associated with PD pathogenesis underlining the value of proteomics in discovering new proteins associated with neurodegenerative disorders.

Four studies have used 2-DE and MS to investigate detergent-soluble proteins in AD brain regions. Although a number of proteins were altered in AD brains, only the increased levels of glyceraldehyde 3-phosphate dehydrogenase (GAPDH) and synaptotagmin I in AD were confirmed, underlining the importance of post-MS verification (Schonberger, Edgar, Kydd, Faull, & Cooper, 2001; Shiozaki et al., 2004).

Several studies have suggested a complex cascade of oxidative damage in neurodegenerative diseases (Lin & Beal, 2006). Proteome analysis of mitochondria, the main generator of superoxide radicals, can indeed provide valuable information on proteins linked to mitochondrial dysfunction. For example, mitochondrial proteomics identified PINK1-interacting proteins, as well as other novel proteins, as potential components of the PINK1 mitophagy pathway, which in turn provided valuable insight into the role of PD-associated proteins in maintaining mitochondrial integrity (Lopez & Melov, 2002; Rodolfo, Ciccosanti, Giacomo, Piacentini, & Fimia, 2010). Proteomic approaches have also been applied in the study of Lewy bodies in PD (Leverenz et al., 2007; Li et al., 2008; Xia et al., 2008). Although valuable, it does not provide accurate details of pathogenic mechanisms as Lewy bodies represent late or terminal events within a pathological cascade. However, SNCA found in Lewy bodies undergo phosphorylation, ubiquitination (Fujiwara et al., 2002), and oxidative nitration, which can be analyzed by a variety of proteomic approaches (Giasson et al., 2000).

Proteomic studies on postmortem cells and tissues that have also added value in confirming existing theories of PD pathogenesis, including mitochondrial impairment, oxidative stress, and changes to the UPS, allowed the discovery of new proteins and their change in characteristics in response to neurodegeneration (Fasano & Lopiano, 2008).

4.3 Proteomic Characterization of Biological Fluids for Biomarker Discovery

Proteins represent attractive biomarker candidates. A successful diagnostic and/or prognostic program relies on accessible protein biomarkers of both normal and disease states particularly if the disease state represents distinct fingerprints of circulating proteins. Indeed, blood represents a prime source of noninvasive biomarkers with broad clinical significance. However, other fluids, such as urine, can also serve as potential biomarker sources. The choice of serum or plasma depends on practical considerations such as ease of sample collection and processing. For example, a target protein may be found bound to coagulation factors favoring plasma as a source or plasma may contain interfering components in which case serum may be the source of choice. Despite this, a uniform blood collection protocol is undoubtedly the most important parameter to ensure consistency in data between different samples.

Cerebrospinal fluid (CSF), although not always easily accessible, represents an excellent source material for fluid proteomics in relation to neurodegenerative disease research, and a variety of CSF proteomic strategies have been employed to study differential protein patterns between disease and control samples. The following sections describe biomarker studies in different neurodegenerative diseases conducted to date.

4.3.1 Parkinson's Disease

Blood and CSF proteomics have been employed to further our understanding of the pathological mechanisms of PD. The rational that RBCs are the major source of DJ-1 prompted the use of whole blood for proteomic analysis at various stages of PD (Zhang et al., 2010). This study using 119 subjects detected the differential expression of 7 DJ-1 isoforms in blood of individuals exposed to 4-hydroxy-2-nonenal (4-HNE), particularly in late-stage of PD. These data were further validated by another whole blood proteomic study carried out by Lin and colleagues using 114 subjects. They suggested the possibility that posttranslationally modified isoforms of DJ-1 from whole blood may act as good biomarkers for late-stage PD rather than total DJ-1 levels (Lin et al., 2012).

A serum proteomic study performed on Chinese PD cohort using 2-DE and MS identified 15 altered proteins, out of which 7 were uniquely associated with PD and involved in antioxidation, lipid metabolism, intracellular transport, cell proliferation, and immunoregulation (Zhao, Xiao, Pu, & Zhong, 2010). Comparative plasma proteomics has also been performed on healthy individuals and PD cohorts using 2-DE and MS and amyloid P has been identified as a possible plasma biomarker for PD (Farooqui, Yang, Rosenberger, & Horrocks, 1997). Comparative CSF proteomics on PD and control individuals has revealed interesting results where proteins such as serum albumin precursor, serum albumin chain-A, hemoglobin β fragment, mutant globin, proline-rich repeat 14 (PRR 14), and serum transferrin N-terminal lobe show altered abundance which in turn could be associated with neuronal dysfunction and the onset and/or progression of PD (Sinha et al., 2009). Redox proteomics has also been employed to identify oxidatively modified proteins from serum and CSF of individuals with AD and PD. This study found that apolipoprotein A-I (ApoA-I) was highly oxidatively modified and its levels were decreased in AD and PD subjects. Combined with the fact that oxidation of ApoA-1 has direct consequences in terms of neurodegeneration, ApoA-I represents a promising diagnostic biomarker as well as a potential target for therapeutic intervention strategies (Keeney et al., 2013). Modified approaches of CSF proteomics have also been developed including (a) MS coupled to magnetic bead-based enrichment of CSF peptides/proteins in combination with advanced statistical analysis (Ishigami et al., 2012), (b) 2D LC-MS/MS coupled to iTRAQ labeling to quantitatively identify altered levels of proteins during different PD stages (Zhang et al., 2010), and (c) LC-MALDI TOF/TOF using isotopic-labeled synthetic peptides as references to identify target peptides without the depletion of abundant proteins from CSF (Pan et al., 2008).

4.3.2 Alzheimer's Disease

CSF protein profiles have also been studied using LC-ESI-MS in response to amyloid precursor proteins and presenilin 1 (PSEN1) gene mutations (Ringman et al., 2012). Although 56 proteins showed altered characteristics between the mutant and nonmutant carriers, significant overlap in CSF protein changes was found between AD groups. In addition to previously identified AD-associated proteins, such as APP, transferrin, α1β-glycoprotein, complement components, afamin precursor, spondin 1, plasminogen, hemopexin, and neuronal pentraxin receptor, the study also identified a number of unique proteins including calsyntenin 3, AMPA (α-amino-3-hydroxy-5-methyl-4-isoxazolepropionic acid) 4 glutamate receptor, CD99 antigen, di-*N*-acetyl-chitobiase, and secreted phosphoprotein 1 (Ringman et al., 2012). SELDI-TOF-MS was recently applied to characterize CSF Aβ peptide isoforms in sporadic PD (Albertini et al., 2012). This study used immunoproteomic approach and identified Aβ1–42 as a potential biomarker for AD. An additional study employed a combination of sample preparation methods, including enrichment of free endogenous peptides produced by proteolysis and peptides noncovalently bound to proteins, to perform a peptidome analysis of postmortem CSF of AD patients and control individuals (Wijte et al., 2012). VGF nerve growth factor inducible precursor and complement C4 precursor were identified in the enriched free peptide fraction, whereas in the protein-bound fraction VGF nerve growth factor inducible precursor and alpha-2-HS-glycoprotein were identified (Wijte et al., 2012). In order to assess the potential of early AD diagnosis through peptidome analysis and to identify additional AD biomarkers, capillary electrophoresis-MS has been developed which reliably detects 1104 low molecular weight peptides in CSF (Jahn et al., 2011).

4.3.3 Huntington's Disease

Only a few proteomic studies have been performed to date in relation to HD. Plasma and CSF proteomics, using comparative 2-DE and MALDI-TOF from HD patients and control individuals, identified clusterin as a potential biomarker for HD (Dalrymple et al., 2007). Huang and colleagues also analyzed the comparative proteome profiles in CSF of six pairs of HD patients and controls using 2-DE and MS (Huang et al., 2011). The CSF proteins prothrombin, apolipoprotein A-IV (Apo A-IV), and haptoglobin were shown to be elevated in HD patients, suggesting their potential role in HD pathogenesis and their potential as biomarkers for HD diagnosis.

4.3.4 Amyotrophic Lateral Sclerosis

As with HD, very few proteomic studies have been performed on ALS. Ranganathan and colleagues used SELDI-TOF-MS to analyze ALS CSF and found that transthyretin and cystatin C levels were decreased and carboxyl-terminal fragment of neuroendocrine protein 7B2 levels were increased in ALS CSF (Ranganathan et al., 2005). Another similar study identified a peptic fragment of neurosecretory protein VGF in ALS CSF (Pasinetti et al., 2006). Using a comparative 2D-DIGE approach on CSF has also suggested fetuin-A and transthyretin as potential biomarkers for ALS (Brettschneider et al., 2010).

4.3.5 Other Neurodegenerative Diseases

Other neurodegenerative diseases such as atypical parkinsonian disorders (APD), including multiple system atrophy and progressive supranuclear palsy, have also been studied by SELDI-TOF-MS. Proteins such as ubiquitin, β2-microglobulin, and two secretogranin 1 fragments were identified that differentiated healthy individuals from individuals with APD (Constantinescu et al., 2010).

5. FROM PROTEOMICS TO CLINICAL SIGNIFICANCE: TARGETING SIGNALING PATHWAYS IN NEURODEGENERATIVE DISEASES

The ultimate aim of clinical proteomic studies is to identify disease-specific biomarkers that can be used as diagnostic and prognostic tools (Hanash, 2003). Kinases, being involved in numerous cellular signaling pathways, represent one of the most important classes of protein in cells and hence have the potential as diagnostic and/or prognostic biomarkers. Targeting kinases in disease has become a hot research topic within the clinical drug development field (Sebolt-Leopold & Herrera, 2004; Williams & Smallridge, 2004). Although somewhat challenging, the analysis of the 518 kinases within the human kinome (representing proteins with more than 20,000 distinct phosphoprotein states) will prove to be paramount in order to further understand the complex network of phosphorylation-based signaling events in both a physiological and pathological state. To date, many proteomic technologies have attempted to study protein phosphorylation using chromatographic enrichment. One of these immobilized metal affinity chromatography (IMAC) using TiO_2 or Fe^{3+} beads is designed to interact with phosphate groups in combination with a cation exchange column (Villen & Gygi, 2008). Phospho-specific Pro-Q(®) Diamond staining, in combination with 2-DE and MS/MS, has been also used to identify phosphoproteins (Stasyk et al., 2005; Steinberg et al., 2003). Within AD tau aggregates, 35 mitotic phosphorylation sites have been identified in the c-Jun N-terminal kinase (JNK)-interacting protein 1(JIP1), using IMAC combined with an LC/MALDI-TOF/ESI-ion trap MS approach. Interestingly, most of the PTMs found were located within the JNK, MAP kinase kinase, and RAC-alpha Ser/Thr protein kinase-binding regions (D'Ambrosio et al., 2006).

Recently, using a combination of 2-DE, Pro-Q(®) Diamond phosphoprotein staining, and LC-MS/MS identified 81 phosphorylated proteins in the substantia nigra and the cortical regions of AD patients as compared to the respective control brain regions. These PTMs will undoubtedly be very useful in establishing a database of potential protein targets of abnormal phosphorylation events in AD brains (Zahid et al., 2012).

Familial PD mutations may result in dysfunctional kinase activities and phosphorylation pathways (Bonifati et al., 2004; Gasser, 2005; Golbe, 2003; Hardy, Cai, Cookson, Gwinn-Hardy, & Singleton, 2006). Protein kinases associated with PD include PINK1 (Valente et al., 2001, 2004), LRRK2 (Funayama et al., 2002; Kumari & Tan, 2009), cyclin-G-associated kinase (GAK) (Dumitriu et al., 2011; Rhodes, Sinsheimer, Bordelon, Bronstein, & Ritz, 2011), and proteins related to phosphorylation signaling pathways such as SNCA (PARK1/4) and DJ-1 (PARK7) (Carballo-Carbajal et al., 2010). Indeed, LRRK2 and several other PD proteins have been associated with MAP and protein kinase B (AKT) kinase signaling pathways (Gloeckner et al., 2010; Hsu, Chan, & Wolozin, 2010). JNK, extracellular signal-regulated kinases (Yuan et al., 1995), and p38, acting downstream of MAP, are particularly important in PD as they play an integral role in neuronal cell death. Potential kinase-related signaling pathway drug targets in PD include protein kinase Cδ, the MLK–JNK signaling cascade, and AKT/protein kinase B signaling cascade, all of which represent kinases implicated in apoptosis (Burke, 2007; Cuny, 2009). Although genetic mutations in SNCA, UCHL1, LRRK2, PINK1, and DJ-1 are clinically associated with PD, their functions are poorly understood (Bonifati, 2010; Fujioka & Wszolek, 2012). The elucidation of their normal function and the pathological mechanisms of the mutated forms have major implications in the development of potential therapeutic intervention strategies for PD.

Although proteomic data for kinases involved in neurodegenerative diseases are limited, future kinomics studies, focusing on understanding kinase signaling pathways, will shed light on biochemical pathways and their cross talk in neurodegeneration.

6. FUTURE PERSPECTIVES: COMBINATORIAL GENOMIC, METABOLOMICS, AND PROTEOMIC ANALYSIS

Patients with neurodegenerative disorders often show extreme heterogeneity in symptom presentation and in the rate of disease progression making diagnosis and prognosis challenging tasks as well as the design of effective intervention strategies (Holmes, 2002; Marras & Lang, 2013). Appropriate neuronal functions, due to metabolic changes, represent a reflection of the complex protein alterations that occur during neurodegenerative disease onset and progression. The combination of proteomic techniques such as affinity purification, the enrichment of proteins of interest, quantitative approaches, and PTM-specific proteomics has started to unravel the cellular networks and mechanisms responsible for the observed heterogeneity. Although proteomics represents a powerful approach to increase our understanding of neurodegeneration, the intrinsic biology underlying neurodegenerative disorders will require the integration of genomics, proteomics, and metabolomics approaches.

ACKNOWLEDGMENTS

Research in our laboratory is funded by The Norwegian Research Council, The Western Norway Health Authority, The Norwegian Parkinson's Disease Association, and The Norwegian Centre for Movement Disorders.

REFERENCES

Acquadro, E., Caron, I., Tortarolo, M., Bucci, E. M., Bendotti, C., & Corpillo, D. (2014). Human SOD1-G93A specific distribution evidenced in murine brain of a transgenic model for amyotrophic lateral sclerosis by MALDI imaging mass spectrometry. *Journal of Proteome Research*, *13*, 1800–1809.

Aggarwal, K., Choe, L. H., & Lee, K. H. (2006). Shotgun proteomics using the iTRAQ isobaric tags. *Briefings in Functional Genomics & Proteomics*, *5*, 112–120.

Aksenov, M. Y., Aksenova, M. V., Carney, J. M., & Butterfield, D. A. (1997). Oxidative modification of glutamine synthetase by amyloid beta peptide. *Free Radical Research*, *27*, 267–281.

Alberio, T., Bossi, A. M., Milli, A., Parma, E., Gariboldi, M. B., Tosi, G., et al. (2010). Proteomic analysis of dopamine and alpha-synuclein interplay in a cellular model of Parkinson's disease pathogenesis. *The FEBS Journal*, *277*, 4909–4919.

Alberio, T., Pippione, A. C., Comi, C., Olgiati, S., Cecconi, D., Zibetti, M., et al. (2012). Dopaminergic therapies modulate the T-CELL proteome of patients with Parkinson's disease. *IUBMB Life*, *64*, 846–852.

Albertini, V., Benussi, L., Paterlini, A., Glionna, M., Prestia, A., Bocchio-Chiavetto, L., et al. (2012). Distinct cerebrospinal fluid amyloid-beta peptide signatures in cognitive decline associated with Alzheimer's disease and schizophrenia. *Electrophoresis*, *33*, 3738–3744.

Albrethsen, J. (2007). Reproducibility in protein profiling by MALDI-TOF mass spectrometry. *Clinical Chemistry*, *53*, 852–858.

Al-Taher, F., Banaszewski, K., Jackson, L., Zweigenbaum, J., Ryu, D., & Cappozzo, J. (2013). Rapid method for the determination of multiple mycotoxins in wines and beers by LC-MS/MS using a stable isotope dilution assay. *Journal of Agricultural and Food Chemistry*, *61*, 2378–2384.

Anderson, N. L., & Anderson, N. G. (1998). Proteome and proteomics: New technologies, new concepts, and new words. *Electrophoresis*, *19*, 1853–1861.

Bantscheff, M., Lemeer, S., Savitski, M. M., & Kuster, B. (2012). Quantitative mass spectrometry in proteomics: Critical review update from 2007 to the present. *Analytical and Bioanalytical Chemistry, 404*, 939–965.

Battistini, S., Giannini, F., Greco, G., Bibbo, G., Ferrera, L., Marini, V., et al. (2005). SOD1 mutations in amyotrophic lateral sclerosis. Results from a multicenter Italian study. *Journal of Neurology, 252*, 782–788.

Bazan, N. G., & Lukiw, W. J. (2002). Cyclooxygenase-2 and presenilin-1 gene expression induced by interleukin-1beta and amyloid beta 42 peptide is potentiated by hypoxia in primary human neural cells. *The Journal of Biological Chemistry, 277*, 30359–30367.

Berciano, J., Infante, J., Mateo, I., & Combarros, O. (2002). Hereditary ataxias and paraplegias: A clinicogenetic review. *Neurología, 17*, 40–51.

Beretta, L. (2007). Proteomics from the clinical perspective: Many hopes and much debate. *Nature Methods, 4*, 785–786.

Bergemalm, D., Forsberg, K., Jonsson, P. A., Graffmo, K. S., Brannstrom, T., Andersen, P. M., et al. (2009). Changes in the spinal cord proteome of an amyotrophic lateral sclerosis murine model determined by differential in-gel electrophoresis. *Molecular & Cellular Proteomics, 8*, 1306–1317.

Bonifati, V. (2010). Shaking the genome: New studies reveal genetic risk for Parkinson's disease. *Lancet Neurology, 9*, 136–138.

Bonifati, V., Oostra, B. A., & Heutink, P. (2004). Unraveling the pathogenesis of Parkinson's disease—The contribution of monogenic forms. *Cellular and Molecular Life Sciences, 61*, 1729–1750.

Boysen, R. I., & Hearn, M. T. (2001). HPLC of peptides and proteins. *Current Protocols in Protein Science, 8*(7), 1–40 (chapter 8).

Brettschneider, J., Lehmensiek, V., Mogel, H., Pfeifle, M., Dorst, J., Hendrich, C., et al. (2010). Proteome analysis reveals candidate markers of disease progression in amyotrophic lateral sclerosis (ALS). *Neuroscience Letters, 468*, 23–27.

Burke, R. E. (2007). Inhibition of mitogen-activated protein kinase and stimulation of Akt kinase signaling pathways: Two approaches with therapeutic potential in the treatment of neurodegenerative disease. *Pharmacology & Therapeutics, 114*, 261–277.

Butterfield, D. A. (2004). Proteomics: A new approach to investigate oxidative stress in Alzheimer's disease brain. *Brain Research, 1000*, 1–7.

Butterfield, D. A., & Dalle-Donne, I. (2012). Redox proteomics. *Antioxidants & Redox Signaling, 17*, 1487–1489.

Cafe-Mendes, C. C., Ferro, E. S., Britto, L. R., & Martins-de-Souza, D. (2014). Using mass spectrometry-based peptidomics to understand the brain and disorders such as Parkinson's disease and schizophrenia. *Current Topics in Medicinal Chemistry, 14*, 369–381.

Campello, L., Esteve-Rudd, J., Bru-Martinez, R., Herrero, M. T., Fernandez-Villalba, E., Cuenca, N., et al. (2013). Alterations in energy metabolism, neuroprotection and visual signal transduction in the retina of Parkinsonian, MPTP-treated monkeys. *PLoS One, 8*, e74439.

Cao, Z., & Robinson, R. A. (2014). Proteome characterization of splenocytes from an Abetapp/ps-1 Alzheimer's disease model. *Proteomics, 14*, 291–297.

Carballo-Carbajal, I., Weber-Endress, S., Rovelli, G., Chan, D., Wolozin, B., Klein, C. L., et al. (2010). Leucine-rich repeat kinase 2 induces alpha-synuclein expression via the extracellular signal-regulated kinase pathway. *Cellular Signalling, 22*, 821–827.

Chang, S. H., Jung, I. S., Han, G. Y., Kim, N. H., Kim, H. J., & Kim, C. W. (2013). Proteomic profiling of brain cortex tissues in a Tau transgenic mouse model of Alzheimer's disease. *Biochemical and Biophysical Research Communications, 430*, 670–675.

Chiang, M. C., Juo, C. G., Chang, H. H., Chen, H. M., Yi, E. C., & Chern, Y. (2007). Systematic uncovering of multiple pathways underlying the pathology of Huntington disease by an acid-cleavable isotope-coded affinity tag approach. *Molecular & Cellular Proteomics*, *6*, 781–797.

Chou, J. L., Shenoy, D. V., Thomas, N., Choudhary, P. K., Laferla, F. M., Goodman, S. R., et al. (2011). Early dysregulation of the mitochondrial proteome in a mouse model of Alzheimer's disease. *Journal of Proteomics*, *74*, 466–479.

Cocciolo, A., Di Domenico, F., Coccia, R., Fiorini, A., Cai, J., Pierce, W. M., et al. (2012). Decreased expression and increased oxidation of plasma haptoglobin in Alzheimer disease: Insights from redox proteomics. *Free Radical Biology & Medicine*, *53*, 1868–1876.

Constantinescu, R., Andreasson, U., Li, S., Podust, V. N., Mattsson, N., Anckarsater, R., et al. (2010). Proteomic profiling of cerebrospinal fluid in parkinsonian disorders. *Parkinsonism & Related Disorders*, *16*, 545–549.

Coppede, F., Mancuso, M., Siciliano, G., Migliore, L., & Murri, L. (2006). Genes and the environment in neurodegeneration. *Bioscience Reports*, *26*, 341–367.

Cornett, D. S., Reyzer, M. L., Chaurand, P., & Caprioli, R. M. (2007). MALDI imaging mass spectrometry: Molecular snapshots of biochemical systems. *Nature Methods*, *4*, 828–833.

Culver, B. P., Savas, J. N., Park, S. K., Choi, J. H., Zheng, S., Zeitlin, S. O., et al. (2012). Proteomic analysis of wild-type and mutant huntingtin-associated proteins in mouse brains identifies unique interactions and involvement in protein synthesis. *The Journal of Biological Chemistry*, *287*, 21599–21614.

Cuny, G. D. (2009). Kinase inhibitors as potential therapeutics for acute and chronic neurodegenerative conditions. *Current Pharmaceutical Design*, *15*, 3919–3939.

Dalrymple, A., Wild, E. J., Joubert, R., Sathasivam, K., Bjorkqvist, M., Petersen, A., et al. (2007). Proteomic profiling of plasma in Huntington's disease reveals neuroinflammatory activation and biomarker candidates. *Journal of Proteome Research*, *6*, 2833–2840.

D'Ambrosio, C., Arena, S., Fulcoli, G., Scheinfeld, M. H., Zhou, D., D'Adamio, L., et al. (2006). Hyperphosphorylation of JNK-interacting protein 1, a protein associated with Alzheimer disease. *Molecular & Cellular Proteomics*, *5*, 97–113.

David, C. (2006). The roles of intracellular protein-degradation pathways in neurodegeneration. *Nature*, *443*, 780–786.

Deschepper, M., Hoogendoorn, B., Brooks, S., Dunnett, S. B., & Jones, L. (2012). Proteomic changes in the brains of Huntington's disease mouse models reflect pathology and implicate mitochondrial changes. *Brain Research Bulletin*, *88*, 210–222.

Di Domenico, F., Sultana, R., Barone, E., Perluigi, M., Cini, C., Mancuso, C., et al. (2011). Quantitative proteomics analysis of phosphorylated proteins in the hippocampus of Alzheimer's disease subjects. *Journal of Proteomics*, *74*, 1091–1103.

Dumitriu, A., Pacheco, C. D., Wilk, J. B., Strathearn, K. E., Latourelle, J. C., Goldwurm, S., et al. (2011). Cyclin-G-associated kinase modifies alpha-synuclein expression levels and toxicity in Parkinson's disease: Results from the GenePD study. *Human Molecular Genetics*, *20*, 1478–1487.

Eng, J. K., McCormack, A. L., & Yates, J. R. (1994). An approach to correlate tandem mass spectral data of peptides with amino acid sequences in a protein database. *Journal of the American Society for Mass Spectrometry*, *5*, 976–989.

Farooqui, A. A., Yang, H. C., Rosenberger, T. A., & Horrocks, L. A. (1997). Phospholipase A2 and its role in brain tissue. *Journal of Neurochemistry*, *69*, 889–901.

Fasano, M., & Lopiano, L. (2008). Alpha-synuclein and Parkinson's disease: A proteomic view. *Expert Review of Proteomics*, *5*, 239–248.

Fenyo, D., & Beavis, R. C. (2003). A method for assessing the statistical significance of mass spectrometry-based protein identifications using general scoring schemes. *Analytical Chemistry*, *75*, 768–774.

Fiehn, O. (2001). Combining genomics, metabolome analysis, and biochemical modelling to understand metabolic networks. *Comparative and Functional Genomics*, *2*, 155–168.

Fountoulakis, M., & Kossida, S. (2006). Proteomics-driven progress in neurodegeneration research. *Electrophoresis*, *27*, 1556–1573.

Fujioka, S., & Wszolek, Z. K. (2012). Update on genetics of parkinsonism. *Neuro-Degenerative Diseases*, *10*, 257–260.

Fujiwara, H., Hasegawa, M., Dohmae, N., Kawashima, A., Masliah, E., Goldberg, M. S., et al. (2002). alpha-Synuclein is phosphorylated in synucleinopathy lesions. *Nature Cell Biology*, *4*, 160–164.

Funayama, M., Hasegawa, K., Kowa, H., Saito, M., Tsuji, S., & Obata, F. (2002). A new locus for Parkinson's disease (PARK8) maps to chromosome 12p11.2-q13.1. *Annals of Neurology*, *51*, 296–301.

Gasser, T. (2005). Genetics of Parkinson's disease. *Current Opinion in Neurology*, *18*, 363–369.

Ghezzi, P., & Bonetto, V. (2003). Redox proteomics: Identification of oxidatively modified proteins. *Proteomics*, *3*, 1145–1153.

Giasson, B. I., Duda, J. E., Murray, I. V., Chen, Q., Souza, J. M., Hurtig, H. I., et al. (2000). Oxidative damage linked to neurodegeneration by selective alpha-synuclein nitration in synucleinopathy lesions. *Science*, *290*, 985–989.

Gloeckner, C. J., Boldt, K., von Zweydorf, F., Helm, S., Wiesent, L., Sarioglu, H., et al. (2010). Phosphopeptide analysis reveals two discrete clusters of phosphorylation in the N-terminus and the Roc domain of the Parkinson-disease associated protein kinase LRRK2. *Journal of Proteome Research*, *9*, 1738–1745.

Golbe, L. I. (2003). Alpha-synuclein and Parkinson's disease. *Advances in Neurology*, *91*, 165–174.

Gorg, A., Weiss, W., & Dunn, M. J. (2004). Current two-dimensional electrophoresis technology for proteomics. *Proteomics*, *4*, 3665–3685.

Guerrera, I. C., & Kleiner, O. (2005). Application of mass spectrometry in proteomics. *Bioscience Reports*, *25*, 71–93.

Gygi, S. P., Rist, B., Gerber, S. A., Turecek, F., Gelb, M. H., & Aebersold, R. (1999). Quantitative analysis of complex protein mixtures using isotope-coded affinity tags. *Nature Biotechnology*, *17*, 994–999.

Hanash, S. (2001). 2-D or not 2-D—Is there a future for 2-D gels in proteomics? Insights from the York proteomics meeting. *Proteomics*, *5*, 635–637.

Hanash, S. (2003). Disease proteomics. *Nature*, *422*, 226–232.

Hands, S. L., & Wyttenbach, A. (2010). Neurotoxic protein oligomerisation associated with polyglutamine diseases. *Acta Neuropathologica*, *120*, 419–437.

Haqqani, A. S., Kelly, J. F., & Stanimirovic, D. B. (2008). Quantitative protein profiling by mass spectrometry using label-free proteomics. *Methods in Molecular Biology*, *439*, 241–256.

Hardy, J., Cai, H., Cookson, M. R., Gwinn-Hardy, K., & Singleton, A. (2006). Genetics of Parkinson's disease and parkinsonism. *Annals of Neurology*, *60*, 389–398.

Hardy, J., & Selkoe, D. J. (2002). The amyloid hypothesis of Alzheimer's disease: Progress and problems on the road to therapeutics. *Science*, *297*, 353–356.

Henkel, A., Müller, K., Lewczuk, P., Müller, T., Marcus, K., Kornhuber, J., et al. (2012). Multidimensional plasma protein separation technique for identification of potential Alzheimer's disease plasma biomarkers: A pilot study. *Journal of Neural Transmission*, *119*, 779–788.

Hensley, K., Hall, N., Subramaniam, R., Cole, P., Harris, M., Aksenov, M., et al. (1995). Brain regional correspondence between Alzheimer's disease histopathology and biomarkers of protein oxidation. *Journal of Neurochemistry*, *65*, 2146–2156.

Ho, C. S., Lam, C. W., Chan, M. H., Cheung, R. C., Law, L. K., Lit, L. C., et al. (2003). Electrospray ionisation mass spectrometry: Principles and clinical applications. *Clinical Biochemist Reviews*, *24*, 3–12.

Holmes, C. (2002). Genotype and phenotype in Alzheimer's disease. *The British Journal of Psychiatry*, *180*, 131–134.

Hong, I., Kang, T., Yoo, Y., Park, R., Lee, J., Lee, S., et al. (2013). Quantitative proteomic analysis of the hippocampus in the 5XFAD mouse model at early stages of Alzheimer's disease pathology. *Journal of Alzheimer's Disease, 36*, 321–334.

Hsu, C. H., Chan, D., & Wolozin, B. (2010). LRRK2 and the stress response: Interaction with MKKs and JNK-interacting proteins. *Neuro-Degenerative Diseases*, 7, 68–75.

Huang, Y. C., Wu, Y. R., Tseng, M. Y., Chen, Y. C., Hsieh, S. Y., & Chen, C. M. (2011). Increased prothrombin, apolipoprotein A-IV, and haptoglobin in the cerebrospinal fluid of patients with Huntington's disease. *PLoS One, 6*, e15809.

Ishigami, N., Tokuda, T., Ikegawa, M., Komori, M., Kasai, T., Kondo, T., et al. (2012). Cerebrospinal fluid proteomic patterns discriminate Parkinson's disease and multiple system atrophy. *Movement Disorders, 27*, 851–857.

Issaq, H. J., Veenstra, T. D., Conrads, T. P., & Felschow, D. (2002). The SELDI-TOF MS approach to proteomics: Protein profiling and biomarker identification. *Biochemical and Biophysical Research Communications, 292*, 587–592.

Jahn, H., Wittke, S., Zurbig, P., Raedler, T. J., Arlt, S., Kellmann, M., et al. (2011). Peptide fingerprinting of Alzheimer's disease in cerebrospinal fluid: Identification and prospective evaluation of new synaptic biomarkers. *PLoS One, 6*, e26540.

Joos, T. O., Schrenk, M., Hopfl, P., Kroger, K., Chowdhury, U., Stoll, D., et al. (2000). A microarray enzyme-linked immunosorbent assay for autoimmune diagnostics. *Electrophoresis, 21*, 2641–2650.

Kawamata, T., Taniguchi, T., Mukai, H., Kitagawa, M., Hashimoto, T., Maeda, K., et al. (1998). A protein kinase, PKN, accumulates in Alzheimer neurofibrillary tangles and associated endoplasmic reticulum-derived vesicles and phosphorylates tau protein. *The Journal of Neuroscience, 18*, 7402–7410.

Keeney, J. T., Swomley, A. M., Forster, S., Harris, J. L., Sultana, R., & Butterfield, D. A. (2013). Apolipoprotein A-I: Insights from redox proteomics for its role in neurodegeneration. *Proteomics. Clinical Applications*, 7, 109–122.

Kim, W., Bennett, E. J., Huttlin, E. L., Guo, A., Li, J., Possemato, A., et al. (2011). Systematic and quantitative assessment of the ubiquitin-modified proteome. *Molecular Cell, 44*, 325–340.

Kislinger, T., & Emili, A. (2003). Going global: Protein expression profiling using shotgun mass spectrometry. *Current Opinion in Molecular Therapeutics, 5*, 285–293.

Kolch, W., & Pitt, A. (2010). Functional proteomics to dissect tyrosine kinase signalling pathways in cancer. *Nature Reviews. Cancer, 10*, 618–629.

Koller, A., Washburn, M. P., Lange, B. M., Andon, N. L., Deciu, C., Haynes, P. A., et al. (2002). Proteomic survey of metabolic pathways in rice. *Proceedings of the National Academy of Sciences of the United States of America, 99*, 11969–11974.

Kumari, U., & Tan, E. K. (2009). LRRK2 in Parkinson's disease: Genetic and clinical studies from patients. *The FEBS Journal, 276*, 6455–6463.

Lauderback, C. M., Hackett, J. M., Huang, F. F., Keller, J. N., Szweda, L. I., Markesbery, W. R., et al. (2001). The glial glutamate transporter, GLT-1, is oxidatively modified by 4-hydroxy-2-nonenal in the Alzheimer's disease brain: The role of Abeta1-42. *Journal of Neurochemistry, 78*, 413–416.

Lehnert, S., Jesse, S., Rist, W., Steinacker, P., Soininen, H., Herukka, S.-K., et al. (2012). iTRAQ and multiple reaction monitoring as proteomic tools for biomarker search in cerebrospinal fluid of patients with Parkinson's disease dementia. *Experimental Neurology, 234*, 499–505.

Le Roch, K. G., Johnson, J. R., Florens, L., Zhou, Y., Santrosyan, A., Grainger, M., et al. (2004). Global analysis of transcript and protein levels across the Plasmodium falciparum life cycle. *Genome Research, 14*, 2308–2318.

Lessner, G., Schmitt, O., Haas, S. J., Mikkat, S., Kreutzer, M., Wree, A., et al. (2010). Differential proteome of the striatum from hemiparkinsonian rats displays vivid structural remodeling processes. *Journal of Proteome Research, 9*, 4671–4687.

Leverenz, J. B., Umar, I., Wang, Q., Montine, T. J., McMillan, P. J., Tsuang, D. W., et al. (2007). Proteomic identification of novel proteins in cortical Lewy bodies. *Brain Pathology*, *17*, 139–145.

Li, Y. H., Wang, J., Zheng, X. L., Zhang, Y. L., Li, X., Yu, S., et al. (2011). Matrix-assisted laser desorption/ionization time-of-flight mass spectrometry combined with magnetic beads for detecting serum protein biomarkers in Parkinson's disease. *European Neurology*, *65*, 105–111.

Li, X., Zhang, Y., Hu, Y., Chang, M., Liu, T., Wang, D., et al. (2008). Chaperone proteins identified from synthetic proteasome inhibitor-induced inclusions in PC12 cells by proteomic analysis. *Acta Biochimica et Biophysica Sinica*, *40*, 406–418.

Licker, V., Cote, M., Lobrinus, J. A., Rodrigo, N., Kovari, E., Hochstrasser, D. F., et al. (2012). Proteomic profiling of the substantia nigra demonstrates CNDP2 overexpression in Parkinson's disease. *Journal of Proteomics*, *75*, 4656–4667.

Lim, Y. A., Rhein, V., Baysang, G., Meier, F., Poljak, A., Raftery, M. J., et al. (2010). Abeta and human amylin share a common toxicity pathway via mitochondrial dysfunction. *Proteomics*, *10*, 1621–1633.

Lin, M. T., & Beal, M. F. (2006). Mitochondrial dysfunction and oxidative stress in neurodegenerative diseases. *Nature*, *443*, 787–795.

Lin, X., Cook, T. J., Zabetian, C. P., Leverenz, J. B., Peskind, E. R., Hu, S. C., et al. (2012). DJ-1 isoforms in whole blood as potential biomarkers of Parkinson disease. *Scientific Reports*, *2*, 954.

Lopez, M. F., & Melov, S. (2002). Applied proteomics: Mitochondrial proteins and effect on function. *Circulation Research*, *90*, 380–389.

Maccioni, R. B., Farias, G., Morales, I., & Navarrete, L. (2010). The revitalized tau hypothesis on Alzheimer's disease. *Archives of Medical Research*, *41*, 226–231.

Manning, G., Whyte, D. B., Martinez, R., Hunter, T., & Sudarsanam, S. (2002). The protein kinase complement of the human genome. *Science*, *298*, 1912–1934.

Marras, C., & Lang, A. (2013). Parkinson's disease subtypes: Lost in translation? *Journal of Neurology, Neurosurgery, and Psychiatry*, *84*, 409–415.

Martin, Z. S., Neugebauer, V., Dineley, K. T., Kayed, R., Zhang, W., Reese, L. C., et al. (2012). alpha-Synuclein oligomers oppose long-term potentiation and impair memory through a calcineurin-dependent mechanism: relevance to human synucleopathic diseases. *Journal of Neurochemistry*, *120*, 440–452.

Matt, P., Fu, Z., Fu, Q., & Van Eyk, J. E. (2008). Biomarker discovery: Proteome fractionation and separation in biological samples. *Physiological Genomics*, *33*, 12–17.

Migliore, L., & Coppede, F. (2002). Genetic and environmental factors in cancer and neurodegenerative diseases. *Mutation Research*, *512*, 135–153.

Mitsui, K., Doi, H., & Nukina, N. (2006). Proteomics of polyglutamine aggregates. *Methods in Enzymology*, *412*, 63–76.

Mudher, A., & Lovestone, S. (2002). Alzheimer's disease—Do tauists and baptists finally shake hands? *Trends in Neurosciences*, *25*, 22–26.

Nilsson, A., Falth, M., Zhang, X., Kultima, K., Skold, K., Svenningsson, P., et al. (2009). Striatal alterations of secretogranin-1, somatostatin, prodynorphin, and cholecystokinin peptides in an experimental mouse model of Parkinson disease. *Molecular & Cellular Proteomics*, *8*, 1094–1104.

Ong, S.-E., Blagoev, B., Kratchmarova, I., Kristensen, D. B., Steen, H., Pandey, A., et al. (2002). Stable isotope labeling by amino acids in cell culture, SILAC, as a simple and accurate approach to expression proteomics. *Molecular & Cellular Proteomics*, *1*, 376–386.

Pan, S., Rush, J., Peskind, E. R., Galasko, D., Chung, K., Quinn, J., et al. (2008). Application of targeted quantitative proteomics analysis in human cerebrospinal fluid using a liquid chromatography matrix-assisted laser desorption/ionization time-of-flight tandem mass spectrometer (LC MALDI TOF/TOF) platform. *Journal of Proteome Research, 7*, 720–730.

Pankratz, N., & Foroud, T. (2007). Genetics of Parkinson disease. *Genetics in Medicine, 9*, 801–811.

Pardo, L. M., & van Duijn, C. M. (2005). In search of genes involved in neurodegenerative disorders. *Mutation Research, 592*, 89–101.

Pasinetti, G. M., Ungar, L. H., Lange, D. J., Yemul, S., Deng, H., Yuan, X., et al. (2006). Identification of potential CSF biomarkers in ALS. *Neurology, 66*, 1218–1222.

Peng, J., Schwartz, D., Elias, J. E., Thoreen, C. C., Cheng, D., Marsischky, G., et al. (2003). A proteomics approach to understanding protein ubiquitination. *Nature Biotechnology, 21*, 921–926.

Periquet, M., Corti, O., Jacquier, S., & Brice, A. (2005). Proteomic analysis of parkin knockout mice: Alterations in energy metabolism, protein handling and synaptic function. *Journal of Neurochemistry, 95*, 1259–1276.

Perluigi, M., Swomley, A. M., & Butterfield, D. A. (2014). Redox proteomics and the dynamic molecular landscape of the aging brain. *Ageing Research Reviews, 13*, 75–89.

Plum, S., Helling, S., Theiss, C., Leite, R. E., May, C., Jacob-Filho, W., et al. (2013). Combined enrichment of neuromelanin granules and synaptosomes from human substantia nigra pars compacta tissue for proteomic analysis. *Journal of Proteomics, 94*, 202–206.

Ranganathan, S., Williams, E., Ganchev, P., Gopalakrishnan, V., Lacomis, D., Urbinelli, L., et al. (2005). Proteomic profiling of cerebrospinal fluid identifies biomarkers for amyotrophic lateral sclerosis. *Journal of Neurochemistry, 95*, 1461–1471.

Rhodes, S. L., Sinsheimer, J. S., Bordelon, Y., Bronstein, J. M., & Ritz, B. (2011). Replication of GWAS associations for GAK and MAPT in Parkinson's disease. *Annals of Human Genetics, 75*, 195–200.

Riederer, I. M., Schiffrin, M., Kovari, E., Bouras, C., & Riederer, B. M. (2009). Ubiquitination and cysteine nitrosylation during aging and Alzheimer's disease. *Brain Research Bulletin, 80*, 233–241.

Rinalducci, S., Murgiano, L., & Zolla, L. (2008). Redox proteomics: Basic principles and future perspectives for the detection of protein oxidation in plants. *Journal of Experimental Botany, 59*, 3781–3801.

Ringman, J. M., Schulman, H., Becker, C., Jones, T., Bai, Y., Immermann, F., et al. (2012). Proteomic changes in cerebrospinal fluid of presymptomatic and affected persons carrying familial Alzheimer disease mutations. *Archives of Neurology, 69*, 96–104.

Rodolfo, C., Ciccosanti, F., Giacomo, G. D., Piacentini, M., & Fimia, G. M. (2010). Proteomic analysis of mitochondrial dysfunction in neurodegenerative diseases. *Expert Review of Proteomics*, 7, 519–542.

Ross, P. L., Huang, Y. N., Marchese, J. N., Williamson, B., Parker, K., Hattan, S., et al. (2004). Multiplexed protein quantitation in *Saccharomyces cerevisiae* using amine-reactive isobaric tagging reagents. *Molecular & Cellular Proteomics, 3*, 1154–1169.

Rudrabhatla, P., Grant, P., Jaffe, H., Strong, M. J., & Pant, H. C. (2010). Quantitative phosphoproteomic analysis of neuronal intermediate filament proteins (NF-M/H) in Alzheimer's disease by iTRAQ. *FASEB Journal, 24*, 4396–4407.

Saez-Valero, J., Barquero, M. S., Marcos, A., McLean, C. A., & Small, D. H. (2000). Altered glycosylation of acetylcholinesterase in lumbar cerebrospinal fluid of patients with Alzheimer's disease. *Journal of Neurology, Neurosurgery, and Psychiatry, 69*, 664–667.

Savino, R., Paduano, S., Preiano, M., & Terracciano, R. (2012). The proteomics big challenge for biomarkers and new drug-targets discovery. *International Journal of Molecular Sciences, 13*, 13926–13948.

Schirmer, E. C., Florens, L., Guan, T., Yates, J. R., 3rd., & Gerace, L. (2003). Nuclear membrane proteins with potential disease links found by subtractive proteomics. *Science, 301*, 1380–1382.

Schonberger, S. J., Edgar, P. F., Kydd, R., Faull, R. L., & Cooper, G. J. (2001). Proteomic analysis of the brain in Alzheimer's disease: Molecular phenotype of a complex disease process. *Proteomics*, *1*, 1519–1528.

Sebolt-Leopold, J. S., & Herrera, R. (2004). Targeting the mitogen-activated protein kinase cascade to treat cancer. *Nature Reviews. Cancer*, *4*, 937–947.

Shiozaki, A., Tsuji, T., Kohno, R., Kawamata, J., Uemura, K., Teraoka, H., et al. (2004). Proteome analysis of brain proteins in Alzheimer's disease: Subproteomics following sequentially extracted protein preparation. *Journal of Alzheimer's Disease*, *6*, 257–268.

Shirasaki, D. I., Greiner, E. R., Al-Ramahi, I., Gray, M., Boontheung, P., Geschwind, D. H., et al. (2012). Network organization of the huntingtin proteomic interactome in mammalian brain. *Neuron*, *75*, 41–57.

Sin, N., Meng, L., Auth, H., & Crews, C. M. (1998). Eponemycin analogues: Syntheses and use as probes of angiogenesis. *Bioorganic & Medicinal Chemistry*, *6*, 1209–1217.

Sinha, A., Srivastava, N., Singh, S., Singh, A. K., Bhushan, S., Shukla, R., et al. (2009). Identification of differentially displayed proteins in cerebrospinal fluid of Parkinson's disease patients: A proteomic approach. *Clinica Chimica Acta*, *400*, 14–20.

Smith, D. J. (2009). Mitochondrial dysfunction in mouse models of Parkinson's disease revealed by transcriptomics and proteomics. *Journal of Bioenergetics and Biomembranes*, *41*, 487–491.

Stasyk, T., Morandell, S., Bakry, R., Feuerstein, I., Huck, C. W., Stecher, G., et al. (2005). Quantitative detection of phosphoproteins by combination of two-dimensional difference gel electrophoresis and phosphospecific fluorescent staining. *Electrophoresis*, *26*, 2850–2854.

Staunton, L., Zweyer, M., Swandulla, D., & Ohlendieck, K. (2012). Mass spectrometry-based proteomic analysis of middle-aged vs. aged vastus lateralis reveals increased levels of carbonic anhydrase isoform 3 in senescent human skeletal muscle. *International Journal of Molecular Medicine*, *30*, 723–733.

Steinberg, T. H., Agnew, B. J., Gee, K. R., Leung, W. Y., Goodman, T., Schulenberg, B., et al. (2003). Global quantitative phosphoprotein analysis using multiplexed proteomics technology. *Proteomics*, *3*, 1128–1144.

Suchkov, S. V., Gnatenko, D. A., Kostiushev, D. S., Krynskii, S. A., & Pal'tsev, M. A. (2013). Proteomics as a fundamental tool for subclinical screening, tests verification and assessment of applied therapy. *Vestnik Rossiĭskoĭ Akademii Meditsinskikh Nauk*, *1*, 65–71.

Sui, X., Ren, X., Huang, P., Li, S., Ma, Q., Ying, M., et al. (2014). Proteomic analysis of serum proteins in triple transgenic Alzheimer's disease mice: Implications for identifying biomarkers for use to screen potential candidate therapeutic drugs for early Alzheimer's disease. *Journal of Alzheimer's Disease*, *40*, 575–586.

Suszynska-Zajczyk, J., Luczak, M., Marczak, L., & Jakubowski, H. (2014). Hyperhomocysteinemia and bleomycin hydrolase modulate the expression of mouse brain proteins involved in neurodegeneration. *Journal of Alzheimer's Disease*, *40*, 713–726.

Sutinen, E. M., Pirttila, T., Anderson, G., Salminen, A., & Ojala, J. O. (2012). Pro-inflammatory interleukin-18 increases Alzheimer's disease-associated amyloid-beta production in human neuron-like cells. *Journal of Neuroinflammation*, *9*, 199.

Takano, M., Yamashita, T., Nagano, K., Otani, M., Maekura, K., Kamada, H., et al. (2013). Proteomic analysis of the hippocampus in Alzheimer's disease model mice by using two-dimensional fluorescence difference in gel electrophoresis. *Neuroscience Letters*, *534*, 85–89.

Tandan, R., & Bradley, W. G. (1985a). Amyotrophic lateral sclerosis: Part 1. Clinical features, pathology, and ethical issues in management. *Annals of Neurology*, *18*, 271–280.

Tandan, R., & Bradley, W. G. (1985b). Amyotrophic lateral sclerosis: Part 2. Etiopathogenesis. *Annals of Neurology*, *18*, 419–431.

Taylor, J. P., Hardy, J., & Fischbeck, K. H. (2002). Toxic proteins in neurodegenerative disease. *Science, 296*, 1991–1995.

Thompson, A., Schafer, J. R., Kuhn, K., Kienle, S., Schwarz, J., Schmidt, G. N., et al. (2003). Tandem mass tags: A novel quantification strategy for comparative analysis of complex protein mixtures by MS/MS. *Analytical Chemistry, 75*, 1895–1904.

Tsuang, D. W., & Bird, T. D. (2002). Genetics of dementia. *The Medical Clinics of North America, 86*, 591–614.

Turk, B. (2006). Targeting proteases: Successes, failures and future prospects. *Nature Reviews. Drug Discovery, 5*, 785–799.

Tyers, M., & Mann, M. (2003). From genomics to proteomics. *Nature, 422*, 193–197.

Ünlü, M., Morgan, M. E., & Minden, J. S. (1997). Difference gel electrophoresis. A single gel method for detecting changes in protein extracts. *Electrophoresis, 18*, 2071–2077.

Valente, E. M., Abou-Sleiman, P. M., Caputo, V., Muqit, M. M., Harvey, K., Gispert, S., et al. (2004). Hereditary early-onset Parkinson's disease caused by mutations in PINK1. *Science, 304*, 1158–1160.

Valente, E. M., Bentivoglio, A. R., Dixon, P. H., Ferraris, A., Ialongo, T., Frontali, M., et al. (2001). Localization of a novel locus for autosomal recessive early-onset parkinsonism, PARK6, on human chromosome 1p35-p36. *American Journal of Human Genetics, 68*, 895–900.

Varga, E., Glauner, T., Koppen, R., Mayer, K., Sulyok, M., Schuhmacher, R., et al. (2012). Stable isotope dilution assay for the accurate determination of mycotoxins in maize by UHPLC-MS/MS. *Analytical and Bioanalytical Chemistry, 402*, 2675–2686.

Vijitruth, R., Liu, M., Choi, D. Y., Nguyen, X. V., Hunter, R. L., & Bing, G. (2006). Cyclooxygenase-2 mediates microglial activation and secondary dopaminergic cell death in the mouse MPTP model of Parkinson's disease. *Journal of Neuroinflammation, 3*, 6.

Villen, J., & Gygi, S. P. (2008). The SCX/IMAC enrichment approach for global phosphorylation analysis by mass spectrometry. *Nature Protocols, 3*, 1630–1638.

Villeneuve, L., Tiede, L. M., Morsey, B., & Fox, H. S. (2013). Quantitative proteomics reveals oxygen-dependent changes in neuronal mitochondria affecting function and sensitivity to rotenone. *Journal of Proteome Research, 12*, 4599–4606.

Washburn, M. P., Wolters, D., & Yates, J. R., 3rd. (2001). Large-scale analysis of the yeast proteome by multidimensional protein identification technology. *Nature Biotechnology, 19*, 242–247.

Westphal, C. H., & Chandra, S. S. (2013). Monomeric synucleins generate membrane curvature. *The Journal of Biological Chemistry, 288*, 1829–1840.

Wijesekera, L. C., & Leigh, P. N. (2009). Amyotrophic lateral sclerosis. *Orphanet Journal of Rare Diseases, 4*, 3.

Wijte, D., McDonnell, L. A., Balog, C. I., Bossers, K., Deelder, A. M., Swaab, D. F., et al. (2012). A novel peptidomics approach to detect markers of Alzheimer's disease in cerebrospinal fluid. *Methods, 56*, 500–507.

Wilkins, M. R., Pasquali, C., Appel, R. D., Ou, K., Golaz, O., Sanchez, J. C., et al. (1996). From proteins to proteomes: Large scale protein identification by two-dimensional electrophoresis and amino acid analysis. *Biotechnology (NY), 14*, 61–65.

Williams, S. F., & Smallridge, R. C. (2004). Targeting the ERK pathway: Novel therapeutics for thyroid cancer. *Current Drug Targets. Immune, Endocrine and Metabolic Disorders, 4*, 199–220.

Wu, C. C., & MacCoss, M. J. (2002). Shotgun proteomics: Tools for the analysis of complex biological systems. *Current Opinion in Molecular Therapeutics, 4*, 242–250.

Xia, Q., Liao, L., Cheng, D., Duong, D. M., Gearing, M., Lah, J. J., et al. (2008). Proteomic identification of novel proteins associated with Lewy bodies. *Frontiers in Bioscience, 13*, 3850–3856.

Xun, Z., Sowell, R. A., Kaufman, T. C., & Clemmer, D. E. (2008). Quantitative proteomics of a presymptomatic A53T alpha-synuclein Drosophila model of Parkinson disease. *Molecular & Cellular Proteomics*, *7*, 1191–1203.

Yang, H., Wittnam, J. L., Zubarev, R. A., & Bayer, T. A. (2013). Shotgun brain proteomics reveals early molecular signature in presymptomatic mouse model of Alzheimer's disease. *Journal of Alzheimer's Disease*, *37*, 297–308.

Yates, J. R., Ruse, C. I., & Nakorchevsky, A. (2009). Proteomics by mass spectrometry: Approaches, advances, and applications. *Annual Review of Biomedical Engineering*, *11*, 49–79.

Ye, Y., & Rape, M. (2009). Building ubiquitin chains: E2 enzymes at work. *Nature Reviews. Molecular Cell Biology*, *10*, 755–764.

Yu, C., & Cohen, L. (2004). Tissue sample preparation—Not the same old grind. *LC-GC Europe*, *17*, 96–111.

Yuan, F., Dellian, M., Fukumura, D., Leunig, M., Berk, D. A., Torchilin, V. P., et al. (1995). Vascular permeability in a human tumor xenograft: Molecular size dependence and cut-off size. *Cancer Research*, *55*, 3752–3756.

Zabel, C., & Klose, J. (2004). Influence of Huntington's disease on the human and mouse proteome. *International Review of Neurobiology*, *61*, 241–283.

Zabel, C., Mao, L., Woodman, B., Rohe, M., Wacker, M. A., Klare, Y., et al. (2009). A large number of protein expression changes occur early in life and precede phenotype onset in a mouse model for huntington disease. *Molecular & Cellular Proteomics*, *8*, 720–734.

Zahid, S., Oellerich, M., Asif, A. R., & Ahmed, N. (2012). Phosphoproteome profiling of substantia nigra and cortex regions of Alzheimer's disease patients. *Journal of Neurochemistry*, *121*, 954–963.

Zareba-Koziol, M., Szwajda, A., Dadlez, M., Wyslouch-Cieszynska, A., & Lalowski, M. (2014). Global analysis of S-nitrosylation sites in the wild type (APP) transgenic mouse brain-clues for synaptic pathology. *Molecular & Cellular Proteomics*, *13*, 2288–2305.

Zhang, X., Zhou, J. Y., Chin, M. H., Schepmoes, A. A., Petyuk, V. A., Weitz, K. K., et al. (2010). Region-specific protein abundance changes in the brain of MPTP-induced Parkinson's disease mouse model. *Journal of Proteome Research*, *9*, 1496–1509.

Zhao, X., Xiao, W. Z., Pu, X. P., & Zhong, L. J. (2010). Proteome analysis of the sera from Chinese Parkinson's disease patients. *Neuroscience Letters*, *479*, 175–179.

Zhou, J. Y., Afjehi-Sadat, L., Asress, S., Duong, D. M., Cudkowicz, M., Glass, J. D., et al. (2010). Galectin-3 is a candidate biomarker for amyotrophic lateral sclerosis: Discovery by a proteomics approach. *Journal of Proteome Research*, *9*, 5133–5141.

Zieske, L. R. (2006). A perspective on the use of iTRAQ reagent technology for protein complex and profiling studies. *Journal of Experimental Botany*, *57*, 1501–1508.

Zilkova, M., Koson, P., & Zilka, N. (2006). The hunt for dying neurons: Insight into the neuronal loss in Alzheimer's disease. *Bratislavské Lekárske Listy*, *107*, 366–373.

CHAPTER THREE

Proteomics Approach to Identify Biomarkers in Neurodegenerative Diseases

Annapurna Nayak*,†, Gregory Salt*, Sunil K. Verma‡, Uday Kishore*,[1]
*Centre for Infection, Immunity and Disease Mechanisms, College of Health and Life Sciences, Brunel University London, Uxbridge, United Kingdom
†Centre for Biotechnology and Bioinformatics, School of Life Sciences, Jawaharlal Nehru Institute for Advanced Studies, Secunderabad, Telangana, India
‡Center for Cellular and Molecular Biology, Hyderabad, Telangana, India
[1]Corresponding author: e-mail addresses: uday.kishore@brunel.ac.uk; ukishore@hotmail.com

Contents

1. Introduction 60
2. Proteomic Methods Utilized in Understanding Neurodegenerative Diseases 61
   2.1 Two-Dimensional Gel Electrophoresis 61
   2.2 Mass Spectrometry 62
3. Alzheimer's Disease 63
   3.1 Pathophysiology 63
4. AD Biomarkers 63
   4.1 Brain Tissue 63
   4.2 Cerebrospinal Fluid 64
   4.3 Plasma 64
5. Parkinson's Disease 68
   5.1 Pathophysiology of Parkinson's Disease 68
6. PD Biomarkers 69
   6.1 Ccerebrospinal Fluid 69
   6.2 Plasma 70
   6.3 Brain Tissue 70
7. Huntington's Disease 71
   7.1 Pathophysiology 71
   7.2 Mitochondrial Dysfunction in HD 72
   7.3 Oxidatively Modified Proteins in HD 73
   7.4 PTM Specific to htt Protein 74
8. Prion Diseases 76
   8.1 Pathophysiology 76
9. Prion Biomarkers 76
   9.1 Brain 76
   9.2 Cerebrospinal Fluid 77

International Review of Neurobiology, Volume 121
ISSN 0074-7742
http://dx.doi.org/10.1016/bs.irn.2015.05.003

10. Amyotrophic Lateral Sclerosis 78
10.1 Pathophysiology 78
11. ALS Biomarkers 79
11.1 Cerebrospinal Fluid 79
11.2 Plasma 79
11.3 Other Biofluids 80
12. Conclusion 80
References 81

Abstract

This chapter examines the use of proteomics in understanding pathogenesis and identifying possible biomarkers in a range of neurodegenerative diseases such as Alzheimer's disease, Parkinson's disease, amyotrophic lateral sclerosis, Huntington's disease, and prion diseases. We have attempted to look at the neuroproteomic approach from a biomarker discovery point of view. Novel biomarkers can pave the way for new therapeutic targets and lead us to a better understanding of the pathogenesis involved in the neurodegenerative diseases.

1. INTRODUCTION

Proteomics is a multidisciplinary approach to the study of the entire range of proteins including their structures, interactions, and posttranslational modifications (PTMs) encoded by an organism's genome under physiological as well as pathological conditions. In pathological conditions, the changes in the protein levels along with their modifications as a result of the pathology, or due to external factors such as toxic agents, can be studied using proteomics. The basic goal of proteomics, in addition to understanding the pathogenesis, is to identify biomarkers and new targets for drug development. The proteomics approach achieves quick, efficient, and reliable protein identification from excisable gel spots, when compared to the arduous process of identifying proteins by staining individual proteins with specific antibodies. Proteomics is particularly useful in the study of pathogenesis of neurodegenerative disorders since these are caused due to misfolded proteins or protein aggregates and the study of these proteins can prove to be vital in developing new biomarkers.

Biomarkers are useful in the assessment of disease progression and extent of proposed therapeutics in the management of the disease. In addition, it is also helpful in the identification of individuals who have possible polymorphisms that might be associated with genetic susceptibility to complex diseases. The main biofluid used in neurodegenerative diseases is cerebrospinal fluid (CSF). Other than CSF, biofluids used for proteomics analysis include serum, plasma, and to a lesser extent, urine and saliva. Serum/plasma samples are much less invasive when collected from patients, but due to the separation of the CSF from blood via the blood–brain barrier, and the fact that the concentrations of proteins in the blood exhibit a range of several orders of magnitude, serum/plasma samples are less useful when looking for biomarkers.

Proteomic analysis of samples comprises two steps: separation of the protein mixture to allow efficient detection of the particular proteins in the mixture through, for instance, traditional two-dimensional gel electrophoresis and multidimensional liquid chromatography and identification of the separated proteins through analytical methods, mainly by mass spectrometry (MS), followed by the use of bioinformatics to analyze the proteins further.

2. PROTEOMIC METHODS UTILIZED IN UNDERSTANDING NEURODEGENERATIVE DISEASES

2.1 Two-Dimensional Gel Electrophoresis

2-DE has been used for over three decades and still remains the main method for separation, comparison, and detection of quantitative changes in biomarker expression levels. Separation of proteins is according to their charge and molecular weight, hence allowing resolution of multiple isoforms and variants of the same proteins, and when coupled with mass spectrometry, identification of individual polypeptide components can be done (Davidsson & Sjogren, 2005). Following separation by 2-DE, the spots detected on the gel are excised and subjected to digestion into peptides, typically with trypsin. An alternative is to subject the digested proteins to separation by single or multidimensional liquid chromatography (LC).

The limitations of 2-DE are the laborious protocols in the separation technology and the fact that few proteins such as acidic/alkaline and hydrophobic proteins are hard to resolve. In addition, quantitative measurements are limited to narrow range and hence most major abundant proteins mask lower abundance proteins. In spite of its limitations, 2-DE gels have been successfully used in the biomarker discovery in neurodegenerative diseases.

2.2 Mass Spectrometry

MS is an efficient reliable technique for detecting the mass of charged proteins. The first step in MS is to ionize and vaporize the analytes (i.e., proteins or their [typically] tryptic peptide fragments) without significant fragmentation. Two commonly utilized approaches are matrix-assisted laser desorption ionization (MALDI) and electrospray ionization (ESI) (Zhang, Keene, Pan, Montine, & Montine, 2008).

MALDI mixes the analyte with one of several different molecules that facilitate ionization and vaporization, while, on the other hand, generation of peptide ions from ESI is considered to be complex. The peptide or protein ions are then introduced into the mass analyzer, such as ion trap, triple quadrupole, time of flight (TOF), and Fourier transform ion cyclotron that are different through their mechanisms of ion separation, mass accuracy, and resolution (Zhang, Keene, et al., 2008).

Some common proteomic techniques use tandem MS (MS–MS) to analyze the daughter or fragment ions of the parent ions whose mass was determined in the first MS dimension. In these collision-induced dissociation tandem MS experiments, the energy applied supports the fragmentation of the peptide bond over amino acid side chains in the parent ion, hence leading to the generation of an ensemble of daughter ions ("b" ions and "y" ions), from which the amino acid sequence can be deduced (Zhang, Keene, et al., 2008). For example, the subproteome in the given sample is first digested to peptides followed by separation by LC, direct ionization by ESI, and then subjected to tandem MS. Another option is to spot the separated peptides directly onto a MALDI plate, followed by tandem MS. Both approaches are much more automated than the traditional 2-DE method, but as there are no gels and fluorescent dyes, quantitative data cannot be derived. To battle this limitation, stable isotope labels such as deuterium or ^{13}C have been introduced into the samples which are then analyzed simultaneously by MS. There are several approaches to the introduction of isotope labels to proteins or peptides; among these are isotope-coded affinity tags (Gygi et al., 1999) and isobaric tags for relative and absolute quantification (iTRAQ) (Ross et al., 2004).

In this chapter, several common as well as rare neurodegenerative diseases will be discussed from a proteomic perspective. The main challenge of working with neurodegenerative diseases is to obtain the samples; i.e., brain samples are the only possible postmortem, CSF can be obtained only through invasive lumbar punctures, and plasma is abundant with immunoglobulins and albumin.

3. ALZHEIMER'S DISEASE

3.1 Pathophysiology

Alzheimer's disease (AD) is a progressive neurodegenerative disease that impairs cognitive functioning and is characterized pathologically by the accumulation of senile plaques and neurofibrillary tangles. AD is the most common cause of dementia affecting >20% of those aged over 80 years. The different forms of AD are early-onset familial AD that affects individuals under the age of 65, familial AD without familial aggregation, and sporadic late-onset AD affecting individuals over the age of 65. Mutations in the APP (Mullan et al., 1992) and presenilins PS1 and PS2 (Citron et al., 1997; Levey et al., 1997) genes are the underlying cause of familial AD, whereas the genetic risk factor for late-onset sporadic AD is a mutation in APOE leading to a mutated isoform of APOE (APOEε4) (Strittmatter & Roses, 1995; Walker et al., 2000). Amyloid plaques, the characteristic feature of AD, consist of extracellular deposition of aggregated Aβ which is a cleavage product of the neuronal APP (Kimberly et al., 2003). Aβ is a peptide of 36–43 amino acids and the two isoforms of Aβ that occur in the amyloid plaques of AD patients are soluble nonfibrillar Aβ1–40 and insoluble fibrillar Aβ1–42. Aβ1–42 has been shown to have a stronger aggregation and deposition potential that Aβ1–40. Aβ1–42 is abundant in the neuritic plaques of AD brains (Gouras et al., 2000). This difference in the pathological potential of the two forms is dictated by the structural and biochemical differences (Roher et al., 1993).

4. AD BIOMARKERS

4.1 Brain Tissue

Proteomics study on AD brain samples, when compared to non-AD nonpathological brain, has been carried out. In one study, 1408 proteins were studied out of which 197 proteins were found to be differentially expressed, i.e., either upregulated or downregulated (Andreev et al., 2012). However, 37 of 197 proteins were reported to be differentially abundant. For instance, some of the proteins reported to be increased in the study were PICALM, PCYT1A, and CAP2 (Andreev et al., 2012). While P-glycoprotein (PGP), sulfotransferase 4A1, protein cGMP-dependent 3′,5′-cyclic phosphodiesterase, and cadherin 23 showed upregulation in this study (Andreev et al., 2012), the same proteins showed underexpression in other studies

(Emilsson, Saetre, & Jazin, 2006; Twine, Janitz, Wilkins, & Janitz, 2011; Vogelgesang et al., 2004). Other dysregulated proteins included ferritin heavy chain, heat-shock protein 70, and those involved in neuronal development (Gozal et al., 2009; Sultana et al., 2007). One novel finding was the differential expression of proteins within the mitochondrial proteome; AUH (methylglutaconyl-CoA hydratase, mitochondrial) was 36-fold overexpressed (FDR = 0.0022), whereas WARS2 (tryptophanyl-tRNA synthetase, mitochondrial) was 18-fold overabundant (Andreev et al., 2012). Although proteomics study using the brain tissue of AD patients is relevant for drug discovery and understanding disease pathogenesis, it seldom contributes to the development of biomarkers unless the same proteins show dysregulation in the CSF and plasma as well as in brain tissue, which can only be obtained after death of the patients. Hence, studies into the CSF and plasma aim to find novel proteins that are more AD specific or more importantly, AD disease progression specific.

4.2 Cerebrospinal Fluid

As the plaques and tangles observed in AD are deposits of Aβ and tau proteins, respectively, it is natural to hypothesize that CSF would contain high amounts of the pathological peptides. However, when CSF was measured for Aβ1–42 levels compared to total nonpathological Aβ, Aβ1–42 levels were significantly lower in AD patients (Motter et al., 1995; Sunderland et al., 2003; Tapiola et al., 2009). This might be due to the preferential deposition of Aβ1–42 in the brain, hence leading to lower levels of Aβ1–42 being secreted into the CSF. However, tau and phosphorylated tau (p-tau) in the same studies were found to be elevated in CSF. Currently, this decrease of Aβ1–42 and increase of tau and p-tau is the accepted observation for AD research and drug discovery. However, there has been some overlap observed in controls and AD patients, thus lowering the chances of the three proteins to be considered as the biomarkers for AD diagnosis. Other studies using CSF from AD patients have reported novel findings such as the identification of the proteins YKL-40, NrCAM, chromogranin A, and carnosinase I (Perrin et al., 2011). Additional proteins that have been identified are listed in Table 1.

4.3 Plasma

Several recent studies have attempted to profile plasma via proteomic analysis in order to narrow down possible biomarkers in AD patients. Over the

Table 1 Proteins of Interest in Various Samples of Neurodegenerative Diseases Identified by Proteomics Analysis

	Brain Samples	CSF	Plasma
Alzheimer's disease	PICALM, PCYT1A, CAP2, P-glycoprotein (PGP), sulfotransferase 4A1, protein cGMP-dependent 3′,5′-cyclic phosphodiesterase, cadherin 23, ferritin heavy chain, heat-shock protein 70, AUH (methylglutaconyl-CoA hydratase), WARS2 (tryptophanyl-tRNAsynthetase), neuronal-specific septin-3, septin-2, septin-5, dihydropteridine reductase, clathrin heavy chain 1	α-2-Macroglobulin α1-Antichymotrypsin α1-Antitrypsin Complement proteins Heat-shock proteins Fibulin-1 Gelsolin Hemopexin Tubulin PEDF TTR Cathepsin D Enolase Creatine kinase Amyloid precursor protein Neuronal pentraxin receptor NrCAM Chromogranin A Apolipoprotein A1, apolipoprotein E, apolipoprotein J, beta-trace, retinol-binding protein, kininogen, alpha-1 antitrypsin, cell cycle progression 8 protein, alpha-1beta glycoprotein NrCAM, YKL-40, chromogranin A, carnosinase I	ApoA1, ApoB100, complement C3, C4b-binding protein, afamin, vitamin D-binding protein precursor, isoform 1 of gelsolin actin regulator, Ig μ chain C region (IGHM), histidine-rich glycoprotein, fibrinogen β and γ chains

Continued

Table 1 Proteins of Interest in Various Samples of Neurodegenerative Diseases Identified by Proteomics Analysis—cont'd

	Brain Samples	CSF	Plasma
Parkinson's disease	Peroxiredoxin II, mitochondrial complex III, ATP synthase D chain, complexin I, profilin, L-type calcium channel delta-subunit, and fatty acid-binding protein, GSH-related detoxification enzymes [glutathione-*S*-transferase Mu3 (GST M3), glutathione-*S*-transferase Pi1 (GST P1), glutathione-*S*-transferase omega1 (GST O1)], peroxiredoxins (PRDX2, PRDX6), cytosolic aldehyde dehydrogenase (ALDH1A1), cellular retinol-binding protein 1 (CRBP1), glia maturation factor β (GMFB), sorcin A, S-adenosylhomocysteine hydrolase (SAHcy hydrolase), seipin, nebulette gamma-glutamyl hydrolase (GGH)	tau, amyloid β42, β2-microglobulin, vitamin D-binding protein, apolipoprotein A-II, apolipoprotein E, brain-derived neurotrophic factor, interleukin-8, complement proteins (factor B, CFH, C3)	C3 (both C3c and C3dg), CFH, factor B

Amyotrophic lateral sclerosis	N/A	Cystatin C, vegetative growth factor (VGF), transthyretin, neuroendocrine protein 7B2	Ferritin, matrix metalloproteinase-9, fibronectin, neurofilament heavy polypeptide, TDP-43, transforming growth factor-β1
Prion diseases	Heme-binding protein 2, integral membrane protein 2B, AFG3-like protein, Ataxin-10, electrogenic sodium bicarbonate cotransporter-1, coronin-2B, formin-2, vacuolar protein sorting-associated protein 51 homolog, GFAP, NSE, MAP-2	Clusterin, heart-type fatty acid-binding protein (H-FABP), Transthyretin, 14-3-3, tau and phosphorylated-tau (phospho.-tau), S-100B, neuron-specific enolase	N/A

years, various proteins were found to be dysregulated in AD patient plasma including α_2-macroglobulin (α_2-M), serum amyloid-p component, ceruloplasmin, and complement factor H (CFH) (Table 1) (Bauer et al., 1991; Hye et al., 2006; Loeffler et al., 1994; Song et al., 2014). These proteins are involved in an array of biological processes involving acute inflammatory response, cholesterol transport, and blood coagulation. One intriguing discovery was the change in the CFH protein in AD plasma (Hye et al., 2006). CFH is a critical negative regulator of the alternative complement pathway and also an inhibitor of the classical complement pathway for a set of C1q ligands (Kishore & Sim, 2012). To reconfirm the significant change, further analysis by immunoblots was carried out that exhibited a specific increase in CFH in AD plasma when compared to other neurodegenerative disorders. This is interesting as CFH has been reported to be colocalized with Aβ plaques (Strohmeyer, Ramirez, Cole, Mueller, & Rogers, 2002; Strohmeyer, Shen, & Rogers, 2000). However, no free CFH outside plaques has been detected in the AD brain or CSF (Strohmeyer et al., 2002). Furthermore, levels of CFH inversely correlate with cognitive decline; i.e., CFH levels are higher in the later stages of the disease (i.e., GDS4 to GDS7 on the Global Deterioration Scale, with stage 7 being the terminally ill stage) (Reisberg, Ferris, de Leon, & Crook, 1982). Another interesting observation is the similarities between AD and another amyloid-related disorder, age-related macular degeneration (AMD), which is characterized by Aβ deposition, and the polymorphic variation in the gene encoding CFH has been associated with AMD (Edwards et al., 2005). Like AD, along with the increase of CFH, an increase of serum amyloid-p component and ceruloplasmin has also been reported in AMD (Mullins, Russell, Anderson, & Hageman, 2000; Newsome et al., 1986). Thus, in addition to proteins involved in metabolic processes, immune pathway molecules are also potentially good biomarkers to deduce the disease progression of AD.

5. PARKINSON'S DISEASE

5.1 Pathophysiology of Parkinson's Disease

Parkinson's disease (PD) is a progressive neurodegenerative disorder that is characterized by severe motor impairment that includes uncontrollable resting tremor, bradykinesia, i.e., slowness of movements, rigidity, and postural instability. PD is the most common neurodegenerative disorder after AD with a prevalence of about 2% among people over 65 years, hence affecting nearly 6 million people worldwide. Classically, there is progressive neurodegeneration in selected brainstem and cerebral nuclei, especially in the substantia nigra pars compacta (SNpc) in the midbrain and the locus ceruleus in the pons. A mild degree of astrocytic and microglial reaction is also observed and the remaining neurons show inclusions that are known as pale bodies and Lewy bodies when located in their cytoplasm, and Lewy neuritis in their neuronal processes (Goedert, Spillantini, Del Tredici, & Braak, 2013). These inclusions are notably composed of α-synuclein (α-SYN) which is a small protein of 140 amino acids that assumes a fibrillary, β-pleated sheet conformation (Spillantini et al., 1997). It binds to other proteins such as the 14-3-3 chaperonin, an anti-apoptotic factor that antagonizes α-SYN and shares 40% sequence conservation with the N-terminal region of α-SYN (Ostrerova et al., 1999). The conformation of α-SYN can range from an unfolded state in solution, to α-helical in the presence of lipid-containing vesicles, and to β-pleated sheet or amyloid structure in the fibrils (Goedert et al., 2013).

There are two different types of PD, sporadic PD and genetic PD, the latter accounting for nearly 15% of all reported PD cases. Thirteen different genes have been associated with genetic PD, for both the autosomal recessive and the autosomal dominant form. Parkin (PARK2), PINK 1 (PARK 6), and DJ-1 (PARK7) are associated with autosomal recessive form and are characterized by early-onset PD, whereas PARK 1 and PARK 4 (α-SYN), UCHL1 (PARK5), and leucine-rich repeat kinase 2 (LRRK1; PARK8) are implicated in autosomal dominant form of PD (Bonifati, Oostra, & Heutink, 2004; Gasser, 2005; Hardy, Cai, Cookson, Gwinn-Hardy, & Singleton, 2006; Lotharius & Brundin, 2002; Zhang, Dawson, & Dawson, 2000).

6. PD BIOMARKERS

6.1 Ccerebrospinal Fluid

Harrington and Merril used a 2-DE approach to detect a single abnormal spot within the Ig light-chain cluster of spots in 75% of CSF samples derived from 20 PD patients (Harrington & Merril, 1984). In addition, other proteins including apolipoprotein A1 and orosomucoid were also identified in the CSF samples. However, these reports have never been confirmed by any further studies.

A recent study used CSF from patients with AD, PD, and other neurological disorders such as dementia with LBs and healthy normal controls and studied changes in the composition of CSF from these individuals (Abdi et al., 2006). Over 1000 proteins were identified, 72 being specific to PD, out of which ceruloplasmin and apolipoprotein H were confirmed by Western blot analysis. The levels of ceruloplasmin correlated with the disease severity. This was challenged by another study that reported an overlap in PD patient and control levels of ceruloplasmin (Hochstrasser et al., 2004; Torsdottir, Sveinbjornsdottir, Kristinsson, Snaedal, & Johannesson, 2006). Yet another recent study used a Luminex-based assay to differentiate eight proteins (tau, amyloid β1–42, β2-macroglobulin, vitamin D-binding protein, apolipoprotein A-II and apolipoprotein E, brain-derived neurotrophic factor (BDNF), and interleukin-8) from the CSF of PD patients (Zhang, Sokal, et al., 2008). However, these observations are yet to be confirmed by other studies.

6.2 Plasma

Another interesting finding, from both a biomarker and drug discovery point of view, is the detection of complement proteins in the serum of PD patients. In this study, sera of 422 patients were studied and differences in the expression of 7 of 9 complement-related proteins were reported (Goldknopf et al., 2006). The sera of both PD and amyotrophic lateral sclerosis (ALS) populations exhibited an increase in C3 (both C3c and C3dg) and CFH, with PD sera showing a specific elevation of factor B. The difference in the complement pathway components suggests the possible role of complement in the pathogenesis of PD by either hampering/mediating immune activation, and hence, inflammation.

6.3 Brain Tissue

Brain samples, especially the SNpc tissue, showed changes in protein expression when compared to normal subjects. In one study, 9 of 44 proteins show significant changes in their levels of expression. For instance, peroxiredoxin II, mitochondrial complex III, ATP synthase D chain, complexin I, profilin, L-type calcium channel delta-subunit, and fatty acid-binding protein were found to be more significantly increased in PD samples than in the control samples (Basso et al., 2003, 2004). Using MALDI-TOF-MS, additional proteins were identified that were important in various stages of the disease pathogenesis. The iron-storing ferritins, especially ferritin H, were found to be increased in SNpc; its expression is considered to be a compensatory mechanism against rising nigral iron levels due to neurodegeneration-induced decrease of iron-storing neuromelanin (Werner, Heyny-von Haussen, Mall, & Wolf, 2008). Furthermore, levels of reduced glutathione (GSH) are decreased in PD and concomitantly GSH-related detoxification enzymes (glutathione-*S*-transferase Mu3 (GST M3), glutathione-*S*-transferase Pi1 (GST P1), and glutathione-*S*-transferase omega1 (GST O1)) have been observed to show increased expression in the PD specimens (Werner et al., 2008). Other proteins that showed substantial increase were peroxiredoxins (PRDX2 and PRDX6), SOD1, and glial activation proteins (glia maturation factor β (GMFB) and sorcin A). Cytosolic aldehyde dehydrogenase (ALDH1A1) is involved in dopamine metabolism and in the detoxification of aldehydes. However, this particular dehydrogenase was found to be significantly reduced. Interestingly, cellular retinol-binding protein 1 (CRBP1), a protein that binds retinol and retinal and then transports them to aldehyde dehydrogenases and hence aids in the metabolism of retinal to retinoic acid, is elevated in PD samples (Krezel et al., 1998). This elevation is likely to be a compensatory mechanism for the loss of the aldehyde dehydrogenase. It has been reported earlier that when retinoid uptake is hampered in Nurr1 knockout mice (where the retinoic acid receptor RXR fails to form heterodimers with the Nurr1 receptor), the mice fail to exhibit development of dopaminergic neurons in the mesencephalon (Zetterstrom et al., 1997).

Over the years, several groups have performed high-throughput quantitative profiling of postmortem PD brain specimens in order to look for novel proteins that might be contributing to disease pathology (Licker et al., 2012, 2014). Some of the proteins (seipin, nebulette, and gamma-glutamyl hydrolase (GGH)) are involved in pathogenic processes such as mitochondrial dysfunction, oxidative stress, or cytoskeleton impairment and/or had functions that have yet to be ascertained (Table 1).

7. HUNTINGTON'S DISEASE

Huntington's disease (HD) is an autosomal dominant neurological disorder that manifests with chorea, dystonia, and behavioral changes. It is progressive, fatal, and can affect mutation carriers at any time in their lives.

7.1 Pathophysiology

The cause is an expansion of a trinucleotide (CAG) repeat section that codes for a polyglutamine (poly-Q) tract at the N-terminus of the huntingtin protein. Huntingtin (htt) is a ubiquitously expressed protein, predicted to be diverse in its roles and interactions. Expansion of the poly-Q segment to a length greater than 35 residues leads to the modification of enzymatic activity upon htt. At the onset of symptoms, cellular abnormalities are detectable despite a lack of obvious neurodegeneration. In these patients, pathological findings upon histological examination of affected neurons of HD patients feature less positive staining for α-tubulin, neurofilament, neuronal fibers, and microtubule-associated protein 2 (DiProspero et al., 2004). This is highly suggestive of dysregulated axonal trafficking and cytoskeletal integrity. Another common feature is protein inclusions, comprising poly-Q/mhtt aggregates, distributed throughout the cytoplasm and nucleus of affected neurons, as seen in HD model mice (Davies et al., 1997). These inclusions are not useful for staging the disease or for prognosis (Mukai et al., 2005). It is becoming evident that toxic intermediate aggregates are the true mediators of neuronal dysfunction and death (Lacor et al., 2007). Affected brain regions of the neurodegenerative processes are primarily the striatum and to a lesser extent, the cortex (Walker et al., 2000).

As the cause of HD is genetic, definitive diagnosis relies on genetic screening rather than analysis of biomarkers from patient samples. As such, proteomic studies of HD investigate primarily the pathophysiological mechanisms involved. However, biomarkers that can predict disease progression can be a useful addition in the prognostic arsenal (Walker, 2007).

The htt protein is structured with the poly-Q stretch at the N-terminus, closely followed by a proline-rich sequence and four HEAT (huntingtin, EF3, PR65-A component of protein phosphatase A, and TOR kinase) sequences (Ehrnhoefer, Sutton, & Hayden, 2011). Sequence prediction analysis of htt has revealed a PEST sequence (proline-, glutamate-, serine-, and threonine-rich sites) at the N-terminus. There are additional PEST sequences: one between HEAT1 and HEAT2, another between HEAT2 and HEAT3, and the last one found in the middle of PEST4 (Warby et al., 2008).

7.2 Mitochondrial Dysfunction in HD

The animal models of HD include chemically induced phenotypes representative of those seen in HD, where inhibitors of mitochondrial succinate dehydrogenase such as 3-nitropropionic acid (3-NP) and malonate are administered to mice. 3-NP is an irreversible inhibitor of succinate dehydrogenase and induces striatal lesions, chorea, dystonia, and reduced cognitive abilities, characteristic of HD (Sorolla et al., 2012). This shows the importance of mitochondrial dysfunction in HD pathogenesis. With mitochondrial metabolism being the main generator of ROS in the cell, dysfunction leads to the release of ROS that oxidatively modifies further proteins and leads to a cascade of protein dysfunction and carbonylation (Butterfield & Dalle-Donne, 2014). One important carbonylation modification involves aconitase, an iron–sulfur protein involved in the Krebs cycle that is particularly susceptible to oxidation due to the $[4Fe–4S]^{2+}$ core in its active site. This has a tendency to release an Fe^{2+} ion when challenged with ROS. Iron is a superoxide generator, and therefore, oxidative modification and subsequent iron release from aconitase modify its activity, impairing the Krebs cycle and ATP synthesis, and amplifies the generation of further ROS. The oxidative stress due to this cascade leads to carbonylation of proteins, often causing their inactivation (Sorolla et al., 2012).

7.3 Oxidatively Modified Proteins in HD

With oxidative stress implicated in HD, many proteins have been found to be oxidatively modified, primarily via carbonylation. Proteins involved in glycolysis that are found to be oxidatively modified in HD include α- and γ-enolase (Perluigi et al., 2005; Sorolla et al., 2010), glyceraldehyde-3-phosphate dehydrogenase, and pyruvate kinase (Sorolla et al., 2010). There are several known energy metabolism proteins oxidatively modified in HD: citrate synthase (Sorolla et al., 2010), aconitase (Sorolla et al., 2008), subunit 2 of cytochrome *b*–*c*1 complex III and the α-subunit of ATP synthase, and mitochondrial creatine kinase (Sorolla et al., 2010).

A number of proteins involved in protection against oxidative stress are themselves oxidatively modified, such as peroxiredoxin 1 (Prx1) and peroxiredoxin 6 (Prx6). These enzymes are also found in higher concentrations in the heavily afflicted striatum of HD patients compared to matched controls and are secreted by microglia and astrocytes. This likely implicates microglial activation in response to oxidative injury in HD (Sorolla et al., 2008). Microglia have been implicated in many neurodegenerative diseases, since their activation into a proinflammatory phenotype by protein aggregates induces an environment of chronic inflammation.

As mutant htt is prone to misfolding and aggregation, proteins that correct aberrant folding are recruited to it, reducing the damage caused by aggregation. Some of these proteins however then suffer oxidative damage, such as heat-shock cognate 71 kDa protein (Hsc71), T-complex protein 1 subunit beta (TCP1 and TRIC) (Sorolla et al., 2010), and heat-shock protein 90 (Hsp90) (Perluigi et al., 2005). Other proteins with such changes include mutant htt itself (Fox et al., 2011), glial fibrillary acidic protein (GFAP) (Sorolla et al., 2008), transitional endoplasmic reticulum ATPase (Sorolla et al., 2010), voltage-dependent anion channel 1 (Perluigi et al., 2005), and *N*(G)-*N*(G)-dimethylarginine dimethylaminohydrolase (Sorolla et al., 2010). Oxidatively modified proteins involved in vitamin B6 metabolism include pyridoxal kinase and α-aminoadipic semialdehyde dehydrogenase (antiquitin 1) (Sorolla et al., 2010). This demonstrably wide variety of oxidatively modified proteins shows the effects of mitochondrial dysfunction, generation of ROS, and inactivation of enzymes as important players in HD (Sorolla et al., 2012). Further to this, it has been reported that oxidatively modified proteins are valuable biomarkers that correlate with HD progression, which could be used to give more accurate prognoses for patients (Tunez et al., 2011).

7.4 PTM Specific to htt Protein

Explored avenues of htt PTM study include looking at palmitoylation, acetylation, phosphorylation, SUMOylation, and ubiquitination (Ehrnhoefer et al., 2011). Palmitoylation, by the addition of a lipid moiety to a molecule, is a method to localize proteins to lipid membranes such as the cell membrane and internal organelles, as well as having other functions. The removal of the palmitoyl group is quick (half-life: 2.5 h) and allows this PTM to play a role in cellular trafficking (Yanai et al., 2006). It has been shown that htt is palmitoylated by the transferase HIP14 (Yanai et al., 2006). Interaction between HIP14 and htt is negatively affected by increased poly-Q length. Introduction of a site-directed mutation that prevents the palmitoylation of htt has been shown to increase neuronal death and confers a propensity of nuclear localization to form insoluble aggregates (Yanai et al., 2006).

C-REB-binding protein is capable of transferring an acetyl group to htt at K444, and this reaction is counteracted by HDAC1. Acetylation of mutant htt leads to the preferential trafficking of the protein to autophagosomes for degradation. Acetylation-resistant mutant htt that was found to accumulate as intracellular aggregates showed increased stability and lessened autophagosomal degradation (Jeong et al., 2009). Thus, HDAC1 may make a potential therapeutic target as its inhibition could lead to greater degradation of mutant htt protein.

The phosphorylation of htt has been described at several sites, such as at the N-terminus in PEST domain 1 (T3, S13, and S16), PEST domain 2 (S421, S434, and S536), close to the third PEST domain (S1181 and S1201), in the intermediary segment between HEAT repeats 3 and 4 (S2067), and in PEST domain 4 (S2653 and S2657) (Ehrnhoefer et al., 2011). Phosphorylation of mutant htt has been identified as a PTM with the potential for lessening disease pathology.

At the N-terminus, where T3 is phosphorylated, this phosphorylation is decreased in correlation with the areas of the CNS with highest pathology, i.e., least phosphorylation in the striatum, followed by the cortex, and then neurons. Interestingly, the levels of T3 phosphorylation are also negatively correlated with the length of the poly-Q stretch. Model experiments have shown that T3 phosphorylation of mhtt leads to greater aggregation (Aiken et al., 2009). S13 and S16 residues are able to be phosphorylated by the IκB kinase (IKK complex). S13 phosphorylates to allow the phoshorylation of S16 and the transfer of these groups are lower for mutant htt than htt (Thompson et al., 2009). These modifications are relevant to disease pathology as the phosphorylation of these residues increases polyubiquitination and subsequent degradation of htt, but this is inhibited once more by the expansion of the poly-Q tract modulating the kinases (Thompson et al., 2009). Interestingly, in the BACHD mouse model of HD, introducing phosphomimetic residues at S13 and S16 ameliorates both the CNS disturbances as well as severely reducing the frequency and size of intracellular aggregates (Gu et al., 2009). Thus, it appears that the N-terminal composition of mutant htt is significant in modulating its affinity for certain PTMs that affect disease development.

The second PEST domain features one of the most studied phosphorylation sites on htt, S421. The phosphorylation of this site is mediated by Akt and SGK (Humbert et al., 2002; Rangone et al., 2004). The phosphorylation of S421 is highest in the least-affected areas (cerebellum) and lowest in the most affected areas (striatum) (Colin et al., 2008). The phosphorylation of S421 plays a role in vesicular transport of BDNF in support of the survival of neurons in the striatum (Baquet, Gorski, & Jones, 2004; Gauthier et al., 2004). Of interest is the fact that BDNF, a powerful neuroprotective factor that can completely protect neurons from poly-Q mhtt-induced death, is supplied to the striatum via the cortex (Baquet et al., 2004). In cells where htt expression is decreased, vesicular transport is slower and more frequently pauses. Htt selectively affects BDNF vesicle transport and thus it is likely that mutant htt negatively regulates this transport, likely through interaction with the HAP/dynein complex (Gauthier et al., 2004), and this causes an increase in the susceptibility of the neurons (particularly in the striatum) to mutant htt fragment-induced neurotoxicity.

8. PRION DISEASES

8.1 Pathophysiology

Prion diseases, also termed transmissible spongiform encephalopathies (TSEs), are a unique group of diseases that can affect both animals and humans alike. Transmission in humans can be genetic (Gerstmann–Sträussler–Scheinker syndrome (GSS) and fatal familial insomnia (FFI)), infectious (variant Creutzfeldt–Jakob disease (vCJD) and kuru), or sporadic CJD (Herbst et al., 2009). Characteristic histopathological features of TSEs are spongiform degeneration, reactive astrocytosis along with accumulation of amyloid in the CNS. The causative agent of TSEs is abnormally accumulated isoforms of the otherwise physiological 33–35 kDa protein $PrP^{cellular}$ (PrP^{c}) (Oesch et al., 1985). The isoform $PrP^{scrapie}$ (PrP^{Sc}) is a result of post-translational conformation conversion and thus has increased β-sheet content in addition to acquiring proteinase K resistance when compared to PrP^{c} which has a predominantly α-helical conformation. Current diagnostic methods fail to detect prion diseases; confirmation is solely during postmortem analyses. Hence, proteomics of plasma or CSF of affected individuals could be of great help to manage the disease at an earlier stage and if infectious, then to control contamination as well.

9. PRION BIOMARKERS

9.1 Brain

Although the accumulation of PrP^{Sc} is the main causative factor in all prion diseases, it has been observed that several other brain proteins involved in vital biological processes and pathways also exhibit changes that might be contributing to the pathogenesis of prion diseases (Shi et al., 2015). Recently, brain samples from cortex and cerebellum were studied from individuals affected with different prion diseases (genetic CJD, sporadic CJD, and FFI) from a proteomics perspective. The approach used was utilization of isobaric peptide tags (iTRAQ) and multidimensional LC-MS/MS to identify and differentiate proteins between prion disease affected and normal brain samples. Some of the upregulated proteins observed in the cortex and cerebellum of the patients, when compared to normal brains, were those involved in the metabolism, protein transport, development of nervous system symptoms such as ataxia, and reorganization of neuronal actin structure (Shi et al., 2015; Table 1). Some of the downregulated proteins include those involved in the actin cytoskeleton assembly and retrograde transport from early and late endosomes to the trans-Golgi network (Shi et al., 2015). Animal studies have shown that prion disease-regulated genes could be found in gene families associated with lysosome organization/biogenesis, immune cell activation/inflammatory response, lipid metabolism, apoptosis, protein biosynthesis/proteolysis, nervous system function/synaptic transmission, and cytoskeleton organization/biogenesis (Rubenstein, 2012). Most of the upregulated genes were associated with proliferation and activation of astrocytes and microglia such as inflammation, cellular movement, lipid metabolism, and lysosomal activity, while the majority of downregulated genes were associated with neuronal and synaptic dysfunction leading to neuronal cell death (Rubenstein, 2012). Interestingly, in human studies, there was no increase of PrP in either the cortex or the cerebellum of the patients; on the contrary, there was a decrease of PrP compared to the normal brains. However, extremely high levels of PrP^{Sc} were detected in the diseases brain samples, as predicted.

9.2 Cerebrospinal Fluid

Currently, few proteins detected in the CSF are being used to diagnose human prion diseases, especially sporadic CJD. For instance, a 30 kDa polypeptide that was identified as a member of the 14-3-3 family of proteins is used in the diagnosis of sporadic CJD, iatrogenic CJD, and genetic CJD. But the limitation of this marker is that it cannot be used in the diagnosis of vCJD, GSS, and FFI. However, the marker is not very reliable as several false-negative and false-positive results have also been reported in sporadic CJD patients (Chapman, McKeel, & Morris, 2000; Green, Knight, Macleod, Lowman, & Will, 2001). Other proteins that are used as diagnostic markers alongside the 14-3-3 protein are β-amyloid, tau-protein, and phosphorylated isoforms, S100b, as well as neuron-specific enolase (Rubenstein, 2012). However, as with the 14-3-3, these proteins are not prion disease specific and are found to be differentially regulated in other neurodegenerative diseases as well. Hence, research to find novel biomarkers that are prion disease specific is paramount at this stage in order to curb and manage the disease at earlier stages. Some of the challenges of finding a single biomarker for prion diseases are the long incubation period of the disease, choice of tissue to be examined, and stage of disease progression.

10. AMYOTROPHIC LATERAL SCLEROSIS

10.1 Pathophysiology

ALS is a disease characterized by muscle weakness and atrophy of both upper and lower motor neurons. It is fatal, and a prognosis, aside from being invariably poor, is currently impossible to accurately predict. Definitive diagnosis of the disease is generally impossible until the patient has progressed to an advanced stage of nervous system decline, as defined by the "El Escorial Criteria." It is an idiopathic disease with cellular features of protein aggregation, oxidative stress, and excitotoxicity (Musaro, 2013). Proteomic studies of ALS are important for both understanding the cause of motor neuron loss in ALS and identifying diagnostic targets.

In familial ALS, approximately 20% of cases are caused by a mutation in the gene coding for superoxide dismutase (SOD1), leading to neurotoxic gain of function (Ripps et al., 1995). There has been intense debate about whether or not this is due to aggregation of mutant SOD1 or other neurotoxic effects such as its incorporation into mitochondria, causing damage and triggering apoptosis (Musaro, 2013). Postmortem CSF analysis by liquid chromatography-tandem mass spectrometry (LC-MS-MS) has shown that GFAP can exhibit significantly more acetylated lysine residues over non-ALS controls. Furthermore, protein capture using anti-acetyl-lysine antibodies precipitates larger GFAP fragments, and larger fragments of GFAP are found in insoluble fractions of CSF analyzed, suggesting that acetylation of GFAP confers degradation resistance and thus prevents proteasome function (Liu et al., 2013). Indeed, it is known that proteasome dysfunction is a feature of neurodegenerative diseases (Dalle-Donne, 2007). Other proteins found to have significantly more acetylated lysine residues in ALS patients included tubulin β-2a, myelin basic protein, and histone 4 (Liu et al., 2013). These proteins are substrates of histone deacetylase (HDAC), which is likely to be dysfunctional in ALS patients. Concordantly, another study found that ALS-mutant SOD1 is capable of modulating HDAC6 activity (Gal et al., 2013), suggesting that the mutant protein may increase acetylation, potentially conferring proteasomal resistance and aiding the aggregation.

11. ALS BIOMARKERS

The search for biomarkers to reliably diagnose ALS has led to the tentative proposal of several candidates, such as cystatin C, vegetative growth factor (VGF), transthyretin, and neuroendocrine protein 7B2 (Ekegren et al., 2008).

11.1 Cerebrospinal Fluid

A comparative study of CSF samples from 100 ALS patients compared to those from patients with multiple sclerosis, AD, other neurological conditions, and healthy controls has been carried out. A SELDI-TOF-MS-derived biomarker panel, generated using the Rule Learner algorithm, demonstrated a 63% sensitivity and 94% specificity (82% overall accuracy) for discriminating ALS from non-ALS controls (Ryberg et al., 2010). The relatively low sensitivity is likely due to the fact that ALS cases are not identical in their pathology. Another significant finding from this study was that while ALS patients were found to have significantly lower native and double-charged transthyretin levels than non-ALS patients, there was an increase in levels of the oxidatively modified Cys-Gly-transthyretin. As ALS is partially characterized by oxidative stress in the CNS, it is likely that this is a result of oxidative damage (Ryberg et al., 2010). It is plausible that oxidatively modified proteins could form a basis for diagnosis of ALS. Another protein potentially useful as a biomarker is cystatin C, a 13.4 kDa protein exhibiting increased CSF levels in ALS patients over controls. This has been proposed in one comparative study that also found reduced transthyretin peaks in ALS patients over non-ALS controls (Ranganathan et al., 2007). CSF analysis seems a good approach for the discovery of important biomarkers for ALS, as it exhibits a reduced abundance of proteins that need to be excluded from the samples before proteomic analysis over serum/plasma. It also resides in proximity to the diseased neurons in the CNS, wherein pathologically altered proteins are likely to be present. However, this is tempered by the invasiveness of the removal procedure (lumbar puncture) and a lower protein content for analysis (Krüger et al., 2013).

11.2 Plasma

Putative biomarkers reported in the serum of ALS patients include increased caspase-1 (as opposed to CSF levels in ALS) (Iłżecka et al., 2001); ferritin, an iron-binding protein that was found to have increased serum levels in ALS patients in a 10-year study (Qureshi et al., 2008); and matrix metalloproteinase-9 (MMP-9), an extracellular matrix-degrading enzyme that degrades fibronectin and collagen and was found to be greatly increased in ALS serum (Beuche et al., 2000). Protein levels found altered in plasma included decreased fibronectin, whose insoluble form is important for cell adhesion (Ono et al., 2000). Fibronectin is a substrate for MMP-9, and decreased levels in plasma may be a result of increased MMP-9 levels in ALS. Plasma proteins found increased in ALS so far include neurofilament heavy polypeptide, TDP-43, and transforming growth factor-β1 (Ganesalingam et al., 2011; Houi et al., 2002; Verstraete et al., 2012).

11.3 Other Biofluids

Urine and saliva, being far removed from the CNS, present less potent diagnostic tools compared to the aforementioned fluids. In urine, type IV collagen was demonstrated to be notably decreased in patients with ALS (Ono et al., 1999). Again, it is possible that this could be partially due to the actions of increased MMP-9 concentration, though this is speculative. In saliva, chromogranin A was detected at higher levels in ALS patients, and in those with terminal ALS (defined as nonfunctioning of two of the following: upper motor neurons, lower motor neurons, spinal cord, plus moderate dysfunction of the third region), it was three times as high as in controls (Obayashi et al., 2008). While this is interesting, early diagnosis is by far the favored result, it is more useful as an indicator of disease progression than for diagnosis.

12. CONCLUSION

In the past couple of decades, new experimental approaches combined with the development of bioinformatics have paved the way forward in the study of various neurodegenerative diseases. The results derived from the use of proteomics experiments have provided insight into the pathogenesis of neurodegeneration with respect to different stages of the diseases and clues in discovering new biomarkers that reflect the central pathogenic mechanisms. By studying postmortem brain samples and CSF and sera from patients, one can study the variations observed in the population affected with the disorders. Hence, further research using proteomics could help in drug discovery, understanding disease mechanisms, detection at earlier stages due to biomarker discovery and hence to a better management of the neurodegenerative diseases.

REFERENCES

Abdi, F., Quinn, J. F., Jankovic, J., McIntosh, M., Leverenz, J. B., Peskind, E., et al. (2006). Detection of biomarkers with a multiplex quantitative proteomic platform in cerebrospinal fluid of patients with neurodegenerative disorders. *Journal of Alzheimer's Disease, 9*(3), 293–348.

Aiken, C. T., Steffan, J. S., Guerrero, C. M., Khashwji, H., Lukacsovich, T., Simmons, D., et al. (2009). Phosphorylation of threonine 3: Implications for Huntingtin aggregation and neurotoxicity. *The Journal of Biological Chemistry, 284*(43), 29427–29436.

Andreev, V. P., Petyuk, V. A., Brewer, H. M., Karpievitch, Y. V., Xie, F., Clarke, J., et al. (2012). Label-free quantitative LC-MS proteomics of Alzheimer's disease and normally aged human brains. *Journal of Proteome Research, 11*(6), 3053–3067.

Baquet, Z. C., Gorski, J. A., & Jones, K. R. (2004). Early striatal dendrite deficits followed by neuron loss with advanced age in the absence of anterograde cortical brain-derived neurotrophic factor. *The Journal of Neuroscience, 24*(17), 4250–4258.

Basso, M., Giraudo, S., Corpillo, D., Bergamasco, B., Lopiano, L., & Fasano, M. (2004). Proteome analysis of human substantia nigra in Parkinson's disease. *Proteomics, 4*(12), 3943–3952.

Basso, M., Giraudo, S., Lopiano, L., Bergamasco, B., Bosticco, E., Cinquepalmi, A., et al. (2003). Proteome analysis of mesencephalic tissues: Evidence for Parkinson's disease. *Neurological Sciences, 24*(3), 155–156.

Bauer, J., Strauss, S., Schreiter-Gasser, U., Ganter, U., Schlegel, P., Witt, I., et al. (1991). Interleukin-6 and alpha-2-macroglobulin indicate an acute-phase state in Alzheimer's disease cortices. *FEBS Letters, 285*(1), 111–114.

Beuche, W., Yushchenko, M., Mader, M., Maliszewska, M., Felgenhauer, K., & Weber, F. (2000). Matrix metalloproteinase-9 is elevated in serum of patients with amyotrophic lateral sclerosis. *NeuroReport, 11*(16), 3419–3422.

Bonifati, V., Oostra, B. A., & Heutink, P. (2004). Unraveling the pathogenesis of Parkinson's disease—The contribution of monogenic forms. *Cellular and Molecular Life Sciences, 61*(14), 1729–1750.

Butterfield, D. A., & Dalle-Donne, I. (2014). Redox proteomics: From protein modifications to cellular dysfunction and disease. *Mass Spectrometry Reviews, 33*(1), 1–6.

Chapman, T., McKeel, D. W., Jr., & Morris, J. C. (2000). Misleading results with the 14-3-3 assay for the diagnosis of Creutzfeldt-Jakob disease. *Neurology, 55*(9), 1396–1397.

Citron, M., Westaway, D., Xia, W., Carlson, G., Diehl, T., Levesque, G., et al. (1997). Mutant presenilins of Alzheimer's disease increase production of 42-residue amyloid beta-protein in both transfected cells and transgenic mice. *Nature Medicine, 3*, 67–72.

Colin, E., Zala, D., Liot, G., Rangone, H., Borrell-Pages, M., Li, X. J., et al. (2008). Huntingtin phosphorylation acts as a molecular switch for anterograde/retrograde transport in neurons. *The EMBO Journal, 27*(15), 2124–2134.

Dalle-Donne, I. (2007). Familial amyotrophic lateral sclerosis (FALS): Emerging hints from redox proteomics. Highlight commentary on: "Redox proteomics analysis of oxidatively modified proteins in G93A-SOD1 transgenic mice—a model of familial amyotrophic lateral sclerosis". *Free Radical Biology & Medicine, 43*(2), 157–159.

Davidsson, P., & Sjogren, M. (2005). The use of proteomics in biomarker discovery in neurodegenerative diseases. *Disease Markers, 21*(2), 81–92.

Davies, S. W., Turmaine, M., Cozens, B. A., DiFiglia, M., Sharp, A. H., Ross, C. A., et al. (1997). Formation of neuronal intranuclear inclusions underlies the neurological dysfunction in mice transgenic for the HD mutation. *Cell, 90*(3), 537–548.

DiProspero, N. A., Chen, E. Y., Charles, V., Plomann, M., Kordower, J. H., & Tagle, D. A. (2004). Early changes in Huntington's disease patient brains involve alterations in cytoskeletal and synaptic elements. *Journal of Neurocytology, 33*(5), 517–533.

Edwards, A. O., Ritter, R., 3rd., Abel, K. J., Manning, A., Panhuysen, C., & Farrer, L. A. (2005). Complement factor H polymorphism and age-related macular degeneration. *Science*, *308*(5720), 421–424.

Ehrnhoefer, D. E., Sutton, L., & Hayden, M. R. (2011). Small changes, big impact: Posttranslational modifications and function of huntingtin in Huntington disease. *The Neuroscientist*, *17*(5), 475–492.

Ekegren, T., Hanrieder, J., & Bergquist, J. (2008). Clinical perspectives of high-resolution mass spectrometry-based proteomics in neuroscience: Exemplified in amyotrophic lateral sclerosis biomarker discovery research. *Journal of Mass Spectrometry: JMS*, *43*(5), 559–571.

Emilsson, L., Saetre, P., & Jazin, E. (2006). Alzheimer's disease: mRNA expression profiles of multiple patients show alterations of genes involved with calcium signaling. *Neurobiology of Disease*, *21*(3), 618–625.

Fox, J. H., Connor, T., Stiles, M., Kama, J., Lu, Z., Dorsey, K., et al. (2011). Cysteine oxidation within N-terminal mutant huntingtin promotes oligomerization and delays clearance of soluble protein. *The Journal of Biological Chemistry*, *286*(20), 18320–18330.

Gal, J., Chen, J., Barnett, K. R., Yang, L., Brumley, E., & Zhu, H. (2013). HDAC6 regulates mutant SOD1 aggregation through two SMIR motifs and tubulin acetylation. *The Journal of Biological Chemistry*, *288*(21), 15035–15045.

Ganesalingam, J., An, J., Shaw, C. E., Shaw, G., Lacomis, D., & Bowser, R. (2011). Combination of neurofilament heavy chain and complement C3 as CSF biomarkers for ALS. *Journal of Neurochemistry*, *117*(3), 528–537.

Gasser, T. (2005). Genetics of Parkinson's disease. *Current Opinion in Neurology*, *18*(4), 363–369.

Gauthier, L. R., Charrin, B. C., Borrell-Pages, M., Dompierre, J. P., Rangone, H., Cordelieres, F. P., et al. (2004). Huntingtin controls neurotrophic support and survival of neurons by enhancing BDNF vesicular transport along microtubules. *Cell*, *118*(1), 127–138.

Goedert, M., Spillantini, M. G., Del Tredici, K., & Braak, H. (2013). 100 years of Lewy pathology. *Nature Reviews. Neurology*, *9*(1), 13–24.

Goldknopf, I. L., Sheta, E. A., Bryson, J., Folsom, B., Wilson, C., Duty, J., et al. (2006). Complement C3c and related protein biomarkers in amyotrophic lateral sclerosis and Parkinson's disease. *Biochemical and Biophysical Research Communications*, *342*(4), 1034–1039.

Gouras, G., Tsai, J., Naslund, J., Vincent, B., Edgar, M., Checler, F., et al. (2000). Intraneuronal Abeta42 accumulation in human brain. *The American Journal of Pathology*, *156*, 15–20.

Gozal, Y. M., Duong, D. M., Gearing, M., Cheng, D., Hanfelt, J. J., Funderburk, C., et al. (2009). Proteomics analysis reveals novel components in the detergent-insoluble subproteome in Alzheimer's disease. *Journal of Proteome Research*, *8*(11), 5069–5079.

Green, A. J., Knight, R. S., Macleod, M. A., Lowman, A., & Will, R. G. (2001). Misleading results with the 14-3-3 assay for the diagnosis of Creutzfeldt-Jakob disease. *Neurology*, *56*(7), 986–987.

Gu, X., Greiner, E. R., Mishra, R., Kodali, R., Osmand, A., Finkbeiner, S., et al. (2009). Serines 13 and 16 are critical determinants of full-length human mutant huntingtin induced disease pathogenesis in HD mice. *Neuron*, *64*(6), 828–840.

Gygi, S. P., Rist, B., Gerber, S. A., Turecek, F., Gelb, M. H., & Aebersold, R. (1999). Quantitative analysis of complex protein mixtures using isotope-coded affinity tags. *Nature Biotechnology*, *17*(10), 994–999.

Hardy, J., Cai, H., Cookson, M. R., Gwinn-Hardy, K., & Singleton, A. (2006). Genetics of Parkinson's disease and parkinsonism. *Annals of Neurology*, *60*(4), 389–398.

Harrington, M. G., & Merril, C. R. (1984). Two-dimensional electrophoresis and "ultrasensitive" silver staining of cerebrospinal fluid proteins in neurological diseases. *Clinical Chemistry*, *30*(12 Pt. 1), 1933–1937.

Herbst, A., McIlwain, S., Schmidt, J. J., Aiken, J. M., Page, C. D., & Li, L. (2009). Prion disease diagnosis by proteomic profiling. *Journal of Proteome Research*, *8*(2), 1030–1036.

Hochstrasser, H., Bauer, P., Walter, U., Behnke, S., Spiegel, J., Csoti, I., et al. (2004). Ceruloplasmin gene variations and substantia nigra hyperechogenicity in Parkinson disease. *Neurology*, *63*(10), 1912–1917.

Houi, K., Kobayashi, T., Kato, S., Mochio, S., & Inoue, K. (2002). Increased plasma TGF-beta1 in patients with amyotrophic lateral sclerosis. *Acta Neurologica Scandinavica*, *106*(5), 299–301.

Humbert, S., Bryson, E. A., Cordelieres, F. P., Connors, N. C., Datta, S. R., Finkbeiner, S., et al. (2002). The IGF-1/Akt pathway is neuroprotective in Huntington's disease and involves Huntingtin phosphorylation by Akt. *Developmental Cell*, *2*(6), 831–837.

Hye, A., Lynham, S., Thambisetty, M., Causevic, M., Campbell, J., Byers, H. L., et al. (2006). Proteome-based plasma biomarkers for Alzheimer's disease. *Brain*, *129*(Pt. 11), 3042–3050.

Iłżecka, J., Stelmasiak, Z., & Dobosz, B. (2001). Interleukin-1beta converting enzyme/Caspase-1 (ICE/Caspase-1) and soluble APO-1/Fas/CD 95 receptor in amyotrophic lateral sclerosis patients. *Acta Neurologica Scandinavica*, *103*(4), 255–258.

Jeong, H., Then, F., Melia, T. J., Jr., Mazzulli, J. R., Cui, L., Savas, J. N., et al. (2009). Acetylation targets mutant huntingtin to autophagosomes for degradation. *Cell*, *137*(1), 60–72.

Kimberly, W. T., Esler, W. P., Ye, W., Ostaszewski, B. L., Gao, J., Diehl, T., et al. (2003). Notch and the amyloid precursor protein are cleaved by similar gamma-secretase(s). *Biochemistry*, *42*(1), 137–144.

Kishore, U., & Sim, R. (2012). Factor H as a regulator of the classical pathway activation. *Immunobiology*, *217*, 162–168.

Krezel, W., Ghyselinck, N., Samad, T. A., Dupe, V., Kastner, P., Borrelli, E., et al. (1998). Impaired locomotion and dopamine signaling in retinoid receptor mutant mice. *Science*, *279*(5352), 863–867.

Krüger, T., Lautenschlager, J., Grosskreutz, J., & Rhode, H. (2013). Proteome analysis of body fluids for amyotrophic lateral sclerosis biomarker discovery. *Proteomics Clinical Applications*, 7(1–2), 123–135.

Lacor, P. N., Buniel, M. C., Furlow, P. W., Clemente, A. S., Velasco, P. T., Wood, M., et al. (2007). Abeta oligomer-induced aberrations in synapse composition, shape, and density provide a molecular basis for loss of connectivity in Alzheimer's disease. *The Journal of Neuroscience*, *27*(4), 796–807.

Levey, A., Heilman, C., Lah, J., Nash, N., Rees, H., Wakai, M., et al. (1997). Presenilin-1 protein expression in familial and sporadic Alzheimer's disease. *Annals of Neurology*, *41*, 742–753.

Licker, V., Cote, M., Lobrinus, J. A., Rodrigo, N., Kovari, E., Hochstrasser, D. F., et al. (2012). Proteomic profiling of the substantia nigra demonstrates CNDP2 overexpression in Parkinson's disease. *Journal of Proteomics*, *75*(15), 4656–4667.

Licker, V., Turck, N., Kovari, E., Burkhardt, K., Cote, M., Surini-Demiri, M., et al. (2014). Proteomic analysis of human substantia nigra identifies novel candidates involved in Parkinson's disease pathogenesis. *Proteomics*, *14*(6), 784–794.

Liu, D., Liu, C., Li, J., Azadzoi, K., Yang, Y., Fei, Z., et al. (2013). Proteomic analysis reveals differentially regulated protein acetylation in human amyotrophic lateral sclerosis spinal cord. *PLoS One*, *8*(12), e80779.

Loeffler, D. A., DeMaggio, A. J., Juneau, P. L., Brickman, C. M., Mashour, G. A., Finkelman, J. H., et al. (1994). Ceruloplasmin is increased in cerebrospinal fluid in

Alzheimer's disease but not Parkinson's disease. *Alzheimer Disease and Associated Disorders*, *8*(3), 190–197.

Lotharius, J., & Brundin, P. (2002). Pathogenesis of Parkinson's disease: Dopamine, vesicles and alpha-synuclein. *Nature Reviews. Neuroscience*, *3*(12), 932–942.

Motter, R., Vigo-Pelfrey, C., Kholodenko, D., Barbour, R., Johnson-Wood, K., Galasko, D., et al. (1995). Reduction of beta-amyloid peptide42 in the cerebrospinal fluid of patients with Alzheimer's disease. *Annals of Neurology*, *38*(4), 643–648.

Mukai, H., Isagawa, T., Goyama, E., Tanaka, S., Bence, N. F., Tamura, A., et al. (2005). Formation of morphologically similar globular aggregates from diverse aggregation-prone proteins in mammalian cells. *Proceedings of the National Academy of Sciences of the United States of America*, *102*(31), 10887–10892.

Mullan, M., Crawford, F., Axelman, K., Houlden, H., Lilius, L., Winblad, B., et al. (1992). A pathogenic mutation for probable Alzheimer's disease in the APP gene at the N-terminus of beta-amyloid. *Nature Genetics*, *1*, 345–347.

Mullins, R. F., Russell, S. R., Anderson, D. H., & Hageman, G. S. (2000). Drusen associated with aging and age-related macular degeneration contain proteins common to extracellular deposits associated with atherosclerosis, elastosis, amyloidosis, and dense deposit disease. *The FASEB Journal*, *14*(7), 835–846.

Musaro, A. (2013). Understanding ALS: New therapeutic approaches. *The FEBS Journal*, *280*(17), 4315–4322.

Newsome, D. A., Swartz, M., Leone, N. C., Hewitt, A. T., Wolford, F., & Miller, E. D. (1986). Macular degeneration and elevated serum ceruloplasmin. *Investigative Ophthalmology & Visual Science*, *27*(12), 1675–1680.

Obayashi, K., Sato, K., Shimazaki, R., Ishikawa, T., Goto, K., Ueyama, H., et al. (2008). Salivary chromogranin A: Useful and quantitative biochemical marker of affective state in patients with amyotrophic lateral sclerosis. *Internal Medicine (Tokyo, Japan)*, *47*(21), 1875–1879.

Oesch, B., Westaway, D., Walchli, M., McKinley, M. P., Kent, S. B., Aebersold, R., et al. (1985). A cellular gene encodes scrapie PrP 27–30 protein. *Cell*, *40*(4), 735–746.

Ono, S., Imai, T., Shimizu, N., Nakayama, M., Mihori, A., Kaneda, K., et al. (2000). Decreased plasma levels of fibronectin in amyotrophic lateral sclerosis. *Acta Neurologica Scandinavica*, *101*(6), 391–394.

Ostrerova, N., Petrucelli, L., Farrer, M., Mehta, N., Choi, P., Hardy, J., et al. (1999). alpha-Synuclein shares physical and functional homology with 14-3-3 proteins. *The Journal of Neuroscience*, *19*(14), 5782–5791.

Perluigi, M., Poon, H. F., Maragos, W., Pierce, W. M., Klein, J. B., Calabrese, V., et al. (2005). Proteomic analysis of protein expression and oxidative modification in r6/2 transgenic mice: A model of Huntington disease. *Molecular & Cellular Proteomics*, *4*(12), 1849–1861.

Perrin, R. J., Craig-Schapiro, R., Malone, J. P., Shah, A. R., Gilmore, P., Davis, A. E., et al. (2011). Identification and validation of novel cerebrospinal fluid biomarkers for staging early Alzheimer's disease. *PLoS One*, *6*(1), e16032.

Qureshi, M., Brown, R. H., Jr., Rogers, J. T., & Cudkowicz, M. E. (2008). Serum ferritin and metal levels as risk factors for amyotrophic lateral sclerosis. *The Open Neurology Journal*, *2*, 51–54.

Ranganathan, S., Nicholl, G. C., Henry, S., Lutka, F., Sathanoori, R., Lacomis, D., et al. (2007). Comparative proteomic profiling of cerebrospinal fluid between living and post mortem ALS and control subjects. *Amyotrophic Lateral Sclerosis: Offical Publication of the World Federation of Neurology Research Group on Motor Neuron Diseases*, *8*(6), 373–379.

Rangone, H., Poizat, G., Troncoso, J., Ross, C. A., MacDonald, M. E., Saudou, F., et al. (2004). The serum- and glucocorticoid-induced kinase SGK inhibits mutant huntingtin-induced toxicity by phosphorylating serine 421 of huntingtin. *The European Journal of Neuroscience*, *19*(2), 273–279.

Reisberg, B., Ferris, S. H., de Leon, M. J., & Crook, T. (1982). The Global Deterioration Scale for assessment of primary degenerative dementia. *The American Journal of Psychiatry*, *139*(9), 1136–1139.

Ripps, M. E., Huntley, G. W., Hof, P. R., Morrison, J. H., & Gordon, J. W. (1995). Transgenic mice expressing an altered murine superoxide dismutase gene provide an animal model of amyotrophic lateral sclerosis. *Proceedings of the National Academy of Sciences of the United States of America*, *92*(3), 689–693.

Roher, A., Lowenson, J., Clarke, S., Wolkow, C., Wang, R., Cotter, R., et al. (1993). Structural alterations in the peptide backbone of β-amyloid core protein may account for its deposition and stability in Alzheimer's disease. *The Journal of Biological Chemistry*, *268*, 3072–3083.

Ross, P. L., Huang, Y. N., Marchese, J. N., Williamson, B., Parker, K., Hattan, S., et al. (2004). Multiplexed protein quantitation in *Saccharomyces cerevisiae* using amine-reactive isobaric tagging reagents. *Molecular & Cellular Proteomics*, *3*(12), 1154–1169.

Rubenstein, R. (2012). Proteomic analysis of prion diseases: Creating clarity or causing confusion? *Electrophoresis*, *33*(24), 3631–3643.

Ryberg, H., An, J., Darko, S., Lustgarten, J. L., Jaffa, M., Gopalakrishnan, V., et al. (2010). Discovery and verification of amyotrophic lateral sclerosis biomarkers by proteomics. *Muscle and Nerve*, *42*(1), 104–111.

Shi, Q., Chen, L. N., Zhang, B. Y., Xiao, K., Zhou, W., Chen, C., et al. (2015). Proteomics analyses for the global proteins in the brain tissues of different human prion diseases. *Molecular & Cellular Proteomics*, *14*(4), 854–869.

Song, F., Poljak, A., Kochan, N. A., Raftery, M., Brodaty, H., Smythe, G. A., et al. (2014). Plasma protein profiling of Mild Cognitive Impairment and Alzheimer's disease using iTRAQ quantitative proteomics. *Proteome Science*, *12*(1), 5.

Sorolla, M. A., Reverter-Branchat, G., Tamarit, J., Ferrer, I., Ros, J., & Cabiscol, E. (2008). Proteomic and oxidative stress analysis in human brain samples of Huntington disease. *Free Radical Biology & Medicine*, *45*(5), 667–678.

Sorolla, M. A., Rodriguez-Colman, M. J., Tamarit, J., Ortega, Z., Lucas, J. J., Ferrer, I., et al. (2010). Protein oxidation in Huntington disease affects energy production and vitamin B6 metabolism. *Free Radical Biology & Medicine*, *49*(4), 612–621.

Sorolla, M. A., Rodriguez-Colman, M. J., Vall-llaura, N., Tamarit, J., Ros, J., & Cabiscol, E. (2012). Protein oxidation in Huntington disease. *Biofactors*, *38*(3), 173–185.

Spillantini, M. G., Schmidt, M. L., Lee, V. M., Trojanowski, J. Q., Jakes, R., & Goedert, M. (1997). Alpha-synuclein in Lewy bodies. *Nature*, *388*(6645), 839–840.

Strittmatter, W., & Roses, A. (1995). Apolipoprotein E and Alzheimer disease. *Proceedings of the National Academy of Sciences of the United States of America*, *92*, 4725–4727.

Strohmeyer, R., Ramirez, M., Cole, G., Mueller, K., & Rogers, J. (2002). Association of factor H of the alternative pathway of complement with agrin and complement receptor 3 in the Alzheimer's disease brain. *Journal of Neuroimmunology*, *131*, 135–146.

Strohmeyer, R., Shen, Y., & Rogers, J. (2000). Detection of complement alternative pathway mRNA and proteins in the Alzheimer's disease brain. *Brain Research. Molecular Brain Research*, *81*, 7–18.

Sultana, R., Boyd-Kimball, D., Cai, J., Pierce, W. M., Klein, J. B., Merchant, M., et al. (2007). Proteomics analysis of the Alzheimer's disease hippocampal proteome. *Journal of Alzheimer's Disease*, *11*(2), 153–164.

Sunderland, T., Linker, G., Mirza, N., Putnam, K. T., Friedman, D. L., Kimmel, L. H., et al. (2003). Decreased beta-amyloid1-42 and increased tau levels in cerebrospinal fluid of patients with Alzheimer disease. *JAMA*, *289*(16), 2094–2103.

Tapiola, T., Alafuzoff, I., Herukka, S. K., Parkkinen, L., Hartikainen, P., Soininen, H., et al. (2009). Cerebrospinal fluid {beta}-amyloid 42 and tau proteins as biomarkers of Alzheimer-type pathologic changes in the brain. *Archives of Neurology*, *66*(3), 382–389.

Thompson, L. M., Aiken, C. T., Kaltenbach, L. S., Agrawal, N., Illes, K., Khoshnan, A., et al. (2009). IKK phosphorylates Huntingtin and targets it for degradation by the proteasome and lysosome. *The Journal of Cell Biology*, *187*(7), 1083–1099.

Torsdottir, G., Sveinbjornsdottir, S., Kristinsson, J., Snaedal, J., & Johannesson, T. (2006). Ceruloplasmin and superoxide dismutase (SOD1) in Parkinson's disease: A follow-up study. *Journal of the Neurological Sciences*, *241*(1–2), 53–58.

Tunez, I., Sanchez-Lopez, F., Aguera, E., Fernandez-Bolanos, R., Sanchez, F. M., & Tasset-Cuevas, I. (2011). Important role of oxidative stress biomarkers in Huntington's disease. *Journal of Medicinal Chemistry*, *54*(15), 5602–5606.

Twine, N. A., Janitz, K., Wilkins, M. R., & Janitz, M. (2011). Whole transcriptome sequencing reveals gene expression and splicing differences in brain regions affected by Alzheimer's disease. *PLoS One*, *6*(1), e16266.

Verstraete, E., Kuiperij, H. B., van Blitterswijk, M. M., Veldink, J. H., Schelhaas, H. J., van den Berg, L. H., et al. (2012). TDP-43 plasma levels are higher in amyotrophic lateral sclerosis. *Amyotrophic Lateral Sclerosis: Offical Publication of the World Federation of Neurology Research Group on Motor Neuron Diseases*, *13*(5), 446–451.

Vogelgesang, S., Warzok, R. W., Cascorbi, I., Kunert-Keil, C., Schroeder, E., Kroemer, H. K., et al. (2004). The role of P-glycoprotein in cerebral amyloid angiopathy; implications for the early pathogenesis of Alzheimer's disease. *Current Alzheimer Research*, *1*(2), 121–125.

Walker, F. O. (2007). Huntington's disease. *Seminars in Neurology*, *27*(2), 143–150.

Walker, L., Pahnke, J., Madauss, M., Vogelgesang, S., Pahnke, A., Herbst, E., et al. (2000). Apolipoprotein E4 promotes the early deposition of Abeta42 and then Abeta40 in the elderly. *Acta Neuropathologica*, *100*, 36–42.

Warby, S. C., Doty, C. N., Graham, R. K., Carroll, J. B., Yang, Y. Z., Singaraja, R. R., et al. (2008). Activated caspase-6 and caspase-6-cleaved fragments of huntingtin specifically colocalize in the nucleus. *Human Molecular Genetics*, *17*(15), 2390–2404.

Werner, C. J., Heyny-von Haussen, R., Mall, G., & Wolf, S. (2008). Proteome analysis of human substantia nigra in Parkinson's disease. *Proteome Science*, *6*, 8.

Yanai, A., Huang, K., Kang, R., Singaraja, R. R., Arstikaitis, P., Gan, L., et al. (2006). Palmitoylation of huntingtin by HIP14 is essential for its trafficking and function. *Nature Neuroscience*, *9*(6), 824–831.

Zetterstrom, R. H., Solomin, L., Jansson, L., Hoffer, B. J., Olson, L., & Perlmann, T. (1997). Dopamine neuron agenesis in Nurr1-deficient mice. *Science*, *276*(5310), 248–250.

Zhang, Y., Dawson, V. L., & Dawson, T. M. (2000). Oxidative stress and genetics in the pathogenesis of Parkinson's disease. *Neurobiology of Disease*, *7*(4), 240–250.

Zhang, J., Keene, C. D., Pan, C., Montine, K. S., & Montine, T. J. (2008). Proteomics of human neurodegenerative diseases. *Journal of Neuropathology and Experimental Neurology*, *67*(10), 923–932.

Zhang, J., Sokal, I., Peskind, E. R., Quinn, J. F., Jankovic, J., Kenney, C., et al. (2008). CSF multianalyte profile distinguishes Alzheimer and Parkinson diseases. *American Journal of Clinical Pathology*, *129*(4), 526–529.

CHAPTER FOUR

Uncovering Neurodegenerative Protein Modifications via Proteomic Profiling

Xavier Gallart-Palau, Aida Serra, Siu Kwan Sze[1]
School of Biological Sciences, Nanyang Technological University, Singapore
[1]Corresponding author: e-mail address: sksze@ntu.edu.sg

Contents

1. Introduction 88
2. Nonenzymatic Degenerative Protein Modifications (DPMs) and Protein Dysfunctions 89
3. Analysis and Characterization of DPMs by Mass Spectrometry-Based Proteomics 92
4. Study of the Brain Proteome by Quantitative Proteomics 94
5. DPMs in Neurodegenerative Diseases 96
   5.1 Oxidation and Nitration 96
   5.2 Racemization and Isomerization 98
   5.3 Carbamylation 99
   5.4 Carbonylation 99
   5.5 Advanced Glycation End-Products 100
6. Protein Dysfunctions and Cognitive Impairment in Neurodegenerative Diseases 100
   6.1 Deamidation of Ion Channel Proteins Na(+)K(+)-ATPase and Dementia 102
   6.2 Deamidation of Synaptic Proteins and Dementia 104
7. Concluding Remarks 106
Acknowledgements 107
References 107

Abstract

Degenerative protein modifications (DPMs) are caused by nonenzymatic chemical reactions that induce changes in protein structure and function which promote disease initiation, pathological progression and also natural aging. These undesirable DPMs include oxidation, carbonylation, carbamylation, glycation, deamidation, isomerization, nitration, and racemization, which impart deleterious structural and functional changes on extracellular matrix proteins and long-lived cell types such as cardiomyocytes and neurons, leading to impaired overall organ function. Despite the obvious clinical importance of understanding DPM biology, the molecular mechanisms that mediate these modifications remain poorly understood largely due to the technical challenges associated with their study. However, recent advances in mass spectrometry-based proteomics technologies now permit global quantitative proteomic profiling of cell lines,

International Review of Neurobiology, Volume 121
ISSN 0074-7742
http://dx.doi.org/10.1016/bs.irn.2015.06.002

animal models, and human clinical samples from a variety of different patient types. These new methods have not only uncovered changes in global protein expression levels but have also identified specific modifications of particular amino acid residues in protein backbones that are associated with disease progression. The nonenzymatic induction of DPMs as revealed by proteomic profiling can help us to better understand the underlying molecular pathology of protein dysfunction in human diseases and natural aging. This chapter discusses recent progress in understanding how proteomic profiling of patient samples derived from the central nervous system can elucidate the DPM biology of human neurodegenerative diseases.

ABBREVIATIONS

AD Alzheimer's disease
AGEs advanced glycation end-products
ALS amyotrophic lateral sclerosis
CNS central nervous system
DPM degenerative protein modification
HD Huntington's disease
iTRAQ isobaric tag for relative and absolute quantitation
NO nitric oxide
PD Parkinson's disease
PTM posttranslational modification
PTN protein tyrosine nitration
ROS reactive oxygen species
VaD vascular dementia

1. INTRODUCTION

The study of protein dysfunction in diseases of the central nervous system (CNS) dates back over 150 years, but progress in understanding the pathology of neurodegenerative diseases has been limited by the lack of suitable methodologies for unraveling their molecular basis. In 1854, the German physician Wirchow used best histochemical methods available at the time to identify wax-like plaques in brain tissues from autopsied individuals. Those unfamiliar features acquired bluish tones when stained with iodine and sulfuric acid, leading Wirchow to conclude that these plaques were composed of "starch." Accordingly, he dubbed these plaques "amyloid," which is derived from the Latin and Greek terms for starch (see Cohen, 1967 for a comprehensive history of amyloidosis). Five years later, Friedrich and Kekule suggested that brain amyloid plaques were in fact composed of proteins rather than carbohydrates, and in so doing they established for the first time the presence of protein dysfunctions in the brain (Sipe & Cohen, 2000). Early in twentieth century, the generalized use of

Congo red (CR) stain allowed the description of the fibril composition of amyloid plaques, as revealed by the contrasting red and the green tones exhibited by stained samples when visualized under polarized optics (Cohen & Calkins, 1959).

From the first discovery of brain amyloidosis in 1854, the presence of protein dysfunctions has been described in almost all neurodegenerative disorders, including Alzheimer's disease (AD), Parkinson's disease (PD), Huntington's disease (HD), and amyotrophic lateral sclerosis (ALS). The amyloid-β peptide and the microtubule-associated protein tau (MAPT) are the key constituents of brain plaques and tangles in AD patients, α-synuclein protein is the main component of brain depositions in PD, huntingtin oligomerization is characteristic of HD brain tissues, and the proteins TDP-43 and SOD1 are typically misfolded in ALS. The tendency of dysfunctional proteins to aggregate and form depositions in neurodegenerative diseases, and the subsequent toxic effects exerted by these fibrillations on brain cell function remain poorly understood. Nonetheless, it is remarkable how the presence and the quantity of plaques in the brain do not correlate with decline in cognitive functions during dementia (Terry et al., 1991).

The initial focus of investigations into the biology of brain amyloidosis was centered on protein oligomers themselves as putative key mediators of the neurodegenerative process. Indeed, this initial hypothesis that protein oligomers initiate proteinopathies has for many years shaped the overall direction of research in neurodegenerative pathologies. However, recent data have indicated that a diverse range of soluble proteins can become dysfunctional and exert toxic effects in brain tissues that surpass those of oligomerized proteins. In spite of this development, there are still only a few reports that have successfully documented the extent of protein dysfunction, modified composition, and altered conformation that occur in neurodegenerative disease. However, as discussed in this review, the most recent technical advances in mass spectrometry-based proteomics have now provided us with the necessary tools to uncover the molecular basis of protein dysfunction in diseases of neurodegeneration.

2. NONENZYMATIC DEGENERATIVE PROTEIN MODIFICATIONS (DPMS) AND PROTEIN DYSFUNCTIONS

Posttranslational modifications (PTMs) regulate protein structure and functions (Mann & Jensen, 2003; Merkley, Metz, Smith, Baynes, & Frizzell, 2014). In healthy cells and tissues, PTMs induced by chemical/

enzymatic modification of protein side chains represent a common mechanism of altering protein function, whereas PTMs generated by nonenzymatic or spontaneous chemical processes are associated with the initiation and progression of disease processes (including inflammation, oxidative stress, proteostasic imbalance, and proteinopathy; Halim et al., 2015; Van Kasteren et al., 2007; Victorino, Mencalha, & Panis, 2014). The protein chemical modifications that take place *in vivo* can occur either spontaneously (e.g., deamidation and racemization), via nonenzymatic chemical reactions (e.g., oxidation and carbamylation), or due to enzymatic activity (Gillery & Jaisson, 2014; Mommen et al., 2014). The nonenzymatic and spontaneous modification of protein side chains are usually associated with loss of function; hence, these have been termed degenerative protein modifications (DPMs) in the current review. Protein DPMs are considered to be a natural consequence of molecular aging, and all known proteins are susceptible to accumulate DPMs during their life span. However, the presence of DPMs in some long-lived host proteins has been reported as associated with pathogenic conditions and diseases (Gillery & Jaisson, 2013, 2014). The DPMs already known to be associated with degenerative diseases and human aging include oxidation, racemization, isomerization, deamidation, nitration, carbonylation, carbamylation, and glycation. Unfortunately, most of these DPMs are irreversible chemical modifications; hence, the defective products of these modifications tend to accumulate in the tissues and can eventually impair overall organ function (Soskic, Groebe, & Schrattenholz, 2008).

DPMs typically exert detrimental effects on the affected proteins. For example, protein oxidation causes severe structural damage due to peptide backbone cleavage, which can lead to the pathological accumulation of misfolded proteins (Berlett & Stadtman, 1997). Oxidation has also been implicated in mediating the pathological effects of smoking, chronic alcoholism, harmful radiation, and ischemia, since all of these environmental stressors promote the massive production of reactive oxygen species (ROS) in host cells (Schuessler & Schilling, 1984). Similarly, protein racemization which changes L-amino acid to D-amino acid occurs during peptide backbone hydrolysis or deamidation; hence, this DPM mainly promotes protein degradation (Powell, Vine, & Crossman, 1992; Ritz, Turzynski, Schutz, Hollmann, & Rochholz, 1996). Racemization can also result from the acquisition of a succinimide ring that forms during deamidation, leading to changes in protein conformation prior to degradation (Clarke, 1987; Fabian, Szendrei, Mantsch, Greenberg, & Otvos, 1994; Lowenson & Clarke, 1988; Orpiszewski & Benson, 1999; Ritz-Timme & Collins,

2002). It has been proposed that racemization can promote protein oligomerization and cellular toxicity when mechanisms of protein quality control are disrupted, e.g., suppression of the ubiquitin–proteasome protein degradation pathway. Formation of succinimide rings can occur due to the rapid deamidation of asparaginyl or glutaminyl residues, but may also arise as a consequence of the slower processes of aspartyl or glutamyl isomerization. In these cases, not only do the resultant protein DPMs promote racemization, but they also lengthen the protein backbone via the addition of atypical amino acid structures known as iso-aspartic acid and iso-glutamic acid (Ritz-Timme & Collins, 2002). Oxidative damage to proteins arising from transcriptional or translational errors has also been proposed as an important precursor of asparagine and glutamine deamidation *in vivo* (Dukan et al., 2000). Deamidation of an asparagine or glutamine residue changes the chemical properties of protein side chains (amide to carboxyl) and in some cases modifies the orientation relative to the rest of the polypeptide chain (Hipkiss, 2006). Accordingly, spontaneous deamidation has been identified as a major source of protein instability and damage (Adav et al., 2014; Gupta & Srivastava, 2004). In the same light, it has been proposed that cellular production of ROS promotes the rearrangement of protein structure and consequently induces "blooming" of isomerized residues (D'angelo et al., 2005). The effects of isomerization on protein structure and function are thought to be largely similar to those of racemization and deamidation. Indeed, both spontaneous deamidation and isomerization can promote protein misfolding and impair basic molecular functions, while also being implicated in the acquisition of toxic characteristics (Desrosiers & Fanélus, 2011).

Among common DPMs, nitration has perhaps been most clearly shown to induce protein dysfunction. The impact of protein nitration on the function of ATP synthase has been studied extensively, and it is now clear that accumulation of nitration products can disrupt the ability of this enzyme to synthesize adenosine triphosphate (ATP) (Abdelmegeed, Jang, Banerjee, Hardwick, & Song, 2013). It has also been demonstrated that accumulation of protein nitrations can delay protein degradation via the ubiquitin–proteasome system (UBS) (Souza et al., 2000), resulting in defective protein aggregation in the affected cells and tissues.

The detrimental effects of oxidative stress on protein function are manifold due to the variety of DPM products this can generate, including oxidized, cleaved, and carbonylated proteins that each exhibit impaired function. For example, the ATPase *SERCA2A* displays reduced ability to hydrolyze ATP following carbonylation of its component lysine or arginine residues (Shao et al., 2011). Similarly, another common DPM known as carbamylation is thought to result from ROS production in host cells, leading to uremia toxicity in end-stage renal disease. All carbamylated proteins identified in previous studies of uremia were marked by total loss of normal biological activity (Erill, Calvo, & Carlos, 1980; Fluckiger, Harmon, Meier, Loo, & Gabbay, 1981; Wang et al., 2007). Accordingly, protein carbamylation has now also been implicated in inflammation, atherogenesis, and smoking-induced chronic diseases (Wang et al., 2007).

In environments that are rich in reduced sugars, proteins may undergo amino acid glycation or form advanced glycation end-products (AGEs) (Brownlee, Cerami, & Vlassara, 1988; Bucala, Makita, Koschinsky, Cerami, & Vlassara, 1993; Ruderman, Williamson, & Brownlee, 1992). Proteins rich in lysine residues are particularly prone to the accumulation of glycations and AGEs (Yan, Chen, et al., 1994), leading to differential modulation of protein function depending on the intra/extracellular location of the affected molecule. AGE-bearing proteins located inside cells tend to oligomerize and represent a source of ROS that can damage multiple other proteins and cellular organelles. Similarly, extracellular matrix proteins that accumulate AGEs can acquire the ability to interact with several important receptors, e.g., the receptor for AGEs (RAGE), a specific type I cell surface receptor or the macrophage scavenger receptor A (SC-A) (Ott et al., 2014). Agonistic interactions of AGEs to their binding receptors activate many cellular pathways, such as the NFκB pathway and microglia proinflammatory pathways, inducing oxidative damage to surrounding cells (Sakurai, Sugioka, & Nakano, 1990; Yan, Schmidt, et al., 1994).

3. ANALYSIS AND CHARACTERIZATION OF DPMs BY MASS SPECTROMETRY-BASED PROTEOMICS

The central dogma of molecular biology states that one gene encodes for a single protein with a defined function, but recent advances achieved in the field of proteomics have challenged this concept by revealing that myriad PTMs can diversify the end-products of individual genes (Cain, Solis, & Cordwell, 2014). In 1959, Fischer, Graves, Crittenden, and Krebs (1959) demonstrated for the first time that certain amino acids in a protein sequence can be modified by the addition of a phosphate group (phosphorylation). This seminal discovery was followed 4 years later by the first report that histones could feature acetylations that would later be determined to exert potent effects on gene expression levels. At the time of these discoveries, standard molecular techniques allowed only the direct analysis of a single protein modification in a purified sample, and detailed analysis of varied PTMs across the whole proteome would remain unreachable until many decades later. Indeed, even after the development of two-dimensional gel electrophoresis, which finally enabled the large-scale study of protein PTMs in human clinical samples, the limited sensitivity of this technique was unable to detect the small molecular weight changes that denote specific DPMs. As a result, the detailed characterization of DPMs in complex mixtures required the evolution of mass spectrometry, which offered substantially increased sensitivity and dynamic ranges of detection than were previously available. Characterization of protein PTMs and/or DPMs by mass spectrometry can be performed using either top-down or bottom-up proteomic methods (Moradian, Kalli, Sweredoski, & Hess, 2014). Top-down proteomics aims to directly analyze intact proteins without prior digestion during sample preparation (Chait, 2006; Han, Jin, Breuker, & McLafferty, 2006; Smith & Kelleher, 2013; Sze, Ge, Oh, & McLafferty, 2002). This approach can detect any type of protein modification arising from chemical modification of the constituent amino acids by measuring mass changes in the precursor ion and resultant fragments during analysis. Although top-down proteomics is perhaps the most suitable approach for the characterization of PTMs/DPMs in complex patient samples, this technique is limited by low protein ionization efficiency as well as the inability of current mass spectrometry technology to efficiently fragment and analyze large protein ions (Dang et al., 2014). At the present time, bottom-up proteomics is a more mature technology that enables proteomic analysis of PTMs/DPMs profiles at the peptide level (Moradian et al., 2014; Tran et al., 2011). This more practical method of DPM analysis uses peptide separation by liquid chromatography followed by tandem mass spectrometry to analyze trypsin-digested peptides. The major limitations of the bottom-up approach are the requirement to perform trypsin digestion prior to analysis, difficulties achieving 100% sequence coverage of proteins (due to loss of improperly digested peptides) and the generation of peptides that are too short/long for robust characterization. In addition, some hydrophilic peptides that exhibit potentially important PTMs/DPMs are not retained in the reversed-phase C_{18} column required to employ the bottom-up approach (Olsen & Mann, 2013). In addition, the identification of relatively low-abundance DPM-bearing peptides in complex patient samples is somewhat limited by the dynamic range of current mass spectrometry instruments. However, various enrichment and chromatographic methods have been developed that enable the enrichment and separation of peptides with specific PTMs/DPMs on a proteome-wide scale (Olsen, Ong, & Mann, 2004; Fig. 1).

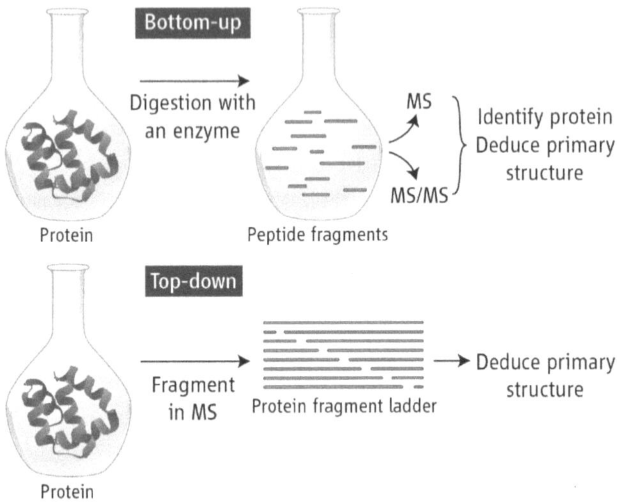

Figure 1 Schematic diagram showing the bottom-up approach (upper) and top-down approach (lower) to protein characterization and proteome analysis. *Adapted from Chait, 2006.*

The characterization of DPMs in complex biological samples is usually restricted by the presence of artifactual DPMs that are generated by the various buffers used in sample preparation. For example, bottom-up proteomic analysis commonly employs a urea buffer which can produce carbamylated protein/peptide artifacts, and optimal trypsin activity at pH 8 can induce experimental protein deamidation. To avoid generating these artifactual DPMs during sample preparation, our group developed a new method of sample processing which substantially improves the accuracy with which protein deamidation can be identified in biomedical samples (Hao & Sze, 2014). This method uses ammonium acetate buffer at pH 6 throughout sample preparation, which maintains sufficient trypsin activity for protein digestion while also restricting the kinetics of asparagine deamidation, thereby minimizing the generation of artifactual deamidation products (Hao, Ren, Alpert, & Sze, 2011; Hao & Sze, 2014).

4. STUDY OF THE BRAIN PROTEOME BY QUANTITATIVE PROTEOMICS

The development of mass spectrometry-based quantitative proteomics, using either label-free approaches or stable isotope labeling (including tandem mass tag and isobaric tag for relative and absolute quantitation [iTRAQ]), presented new opportunities to analyze and characterize the whole-brain proteome in both health and disease. Our group pioneered the application of these techniques to the study of changes in the human brain proteome in both ischemic stroke (Datta, Akatsu, Heese, & Sze, 2013; Datta et al., 2011, 2010) and vascular dementia (VaD) (Adav et al., 2014; Datta, Qian, et al., 2014; Gallart-Palau et al., 2015). For the VaD study, we pooled postmortem brain tissues from 10 healthy subjects and 10 VaD patients who were matched for postmortem delay and also assessed neuropathologically prior to inclusion. The region of the brain used for these studies was the middle temporal cortex, area BA21 according to the Brodmann coordinates. We then used iTRAQ technology together with a multiplexing strategy which detected the fragmented ions of mass tags that generated a reporter ion in one of four different channels on the tandem mass spectra (Ong & Mann, 2005; Ross et al., 2004; Fig. 2).

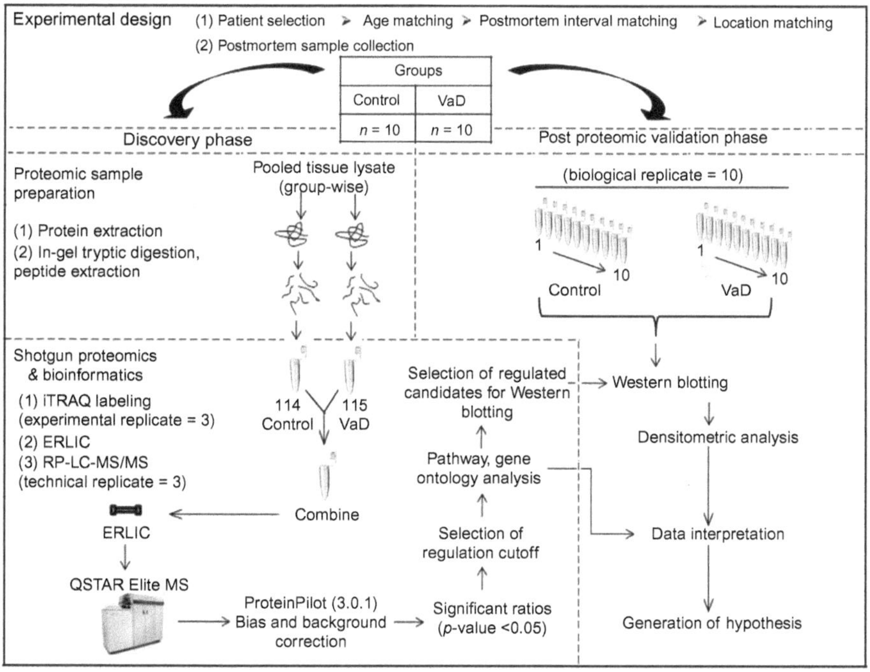

Figure 2 Workflow of a typical neuroproteomics experiment. The experimental design includes collection and selection of demographic-matched case–control samples. The samples are first used for proteomic profiling in the discovery phase and then for biochemical study in the validation phase. The quantitative proteomic study can be performed using stable isotope labeling (as showed in the figure) or label-free methods. After data analysis and data mining, protein targets/candidates associated with disease are shortlisted for biochemical validation and generation of testable hypotheses for further functional studies. *Adapted from Datta, Qian, et al. (2014).*

Use of quantitative proteomics to profile the human brain proteome presents clear advantages, since this discovery-driven approach can generate unforeseen insight into disease pathology which is likely to be overlooked by classical hypothesis-driven research (Lubec, Krapfenbauer, & Fountoulakis, 2003). The discovery-driven approach also has the advantage of being unbiased by prior assumptions, thus allowing the analysis and comparison between experimental conditions and data from thousands of individual peptides and proteins. The application of proteomics is usually divided into two primary fields, known as *profiling proteomics* and *functional proteomics* (Choudhary & Grant, 2004). Profiling proteomics consists of whole-proteome analyses and determination of the changes that occur under specific experimental conditions, whereas functional proteomics characterizes the activity of the studied proteins, the interaction of these proteins with other molecules, and the PTMs that influence their properties. Application of iTRAQ quantitative proteomics to profiling of the brain proteome can provide data for both types of analysis, thereby maximizing potential to identify therapeutic targets among the underlying molecular mediators of neurodegenerative diseases (Adav et al., 2014; Craft, Chen, & Nairn, 2013; Datta, Qian, et al., 2014; Gallart-Palau et al., 2015). Despite this ability to yield useful data on the molecular pathology of neurodegeneration, use of these techniques is complicated by the extreme complexity of CNS tissues, which restricts the characterization of low-abundance proteins (Choudhary & Grant, 2004). Another drawback is the relative expense of the stable isotope labeling reagents and instrumentation required, which necessitates the pooling of samples from multiple individual subjects (Datta, Chen, & Sze, 2014; Datta et al., 2011, 2010; Datta, Qian, et al., 2014; Dutta, Yan, Lim, Tam, & Sze, 2014). Yet another challenge associated with the use of quantitative proteomics to analyze brain tissues is the inability to assign the identified and characterized proteins to specific cell types in the affected tissues. Indeed, it is almost impossible to associate certain levels of an identified protein with a specific subpopulation of neurons (Craft et al., 2013). Finally, the implementation of robust statistics and the application of complementary validation methods are good strategies for maximizing the potential of proteomics to uncover pathological modifications of the brain proteome.

5. DPMs IN NEURODEGENERATIVE DISEASES

5.1 Oxidation and Nitration

Massive production of ROS in neurons and glial cells is a common hallmark of neurodegenerative diseases (Halliwell, 2006; Lim, Chua, Palau, &

Yao, 2014), and accordingly there are a vast number of brain tissue proteins that exhibit modulation by oxidative stress (Bayes & Grant, 2009). In the case of AD, the amyloid β-peptide 1–42 has been found to be highly oxidized in human brain and can promote the oxidation of multiple other critical proteins both *in vitro* and *in vivo*, which may represent a key contributing factor to disease progression in AD (Butterfield, 2003; Butterfield, Drake, Pocernich, & Castegna, 2001; Butterfield & Lauderback, 2002; Selkoe, 2001; Varadarajan, Yatin, Aksenova, & Butterfield, 2000). Proteins involved in neuronal excitotoxicity such as glutamine synthetase and glutamate transporter EAAT2 have also been found to exhibit significant oxidation in the human brain during AD, leading to significant loss of function in both proteins and the accumulation of glutamate in the extracellular domains (Aksenov, Aksenova, Butterfield, Geddes, & Markesbery, 2001; Masliah, Alford, Deteresa, Mallory, & Hansen, 1996). The chaperone protein Hsp60 and the tubulin β-chain have similarly been revealed by proteomics to exhibit oxidation due to the action of the amyloid β-peptide 1–42 (Boyd-Kimball et al., 2005), and deterioration of these proteins has detrimental effects on cellular dynamics and proteostasis within the neurons of affected brain regions.

It is still unknown whether oxidative stress itself can trigger neurodegeneration or simply represents a by-product of the complex neurodegenerative process. The pathology of PD begins at clusters of dopaminergic neurons in the midbrain nigrostriatal system, and it has previously been hypothesized that the dopamine molecule would have a predisposition to become oxidized. Indeed, normal enzymatic degradation of dopamine in dopaminergic neurons is known to generate hydrogen peroxide as a side product (Adams, Murrill, McCreery, Blank, & Karolczak, 1972; Lavoie, Ostaszewski, Weihofen, Schlossmacher, & Selkoe, 2005). The protein α-synuclein is highly oxidized and accumulates in oligomerized forms in PD brain tissues (Kruger et al., 1998), and the oxidized form has been detected in multiple different brain regions in patients with sporadic PD (Jenner & Olanow, 1998), suggesting that accumulation of this protein is a feature of vulnerable basal ganglia neurons.

The impact of oxidation on CNS protein function has been studied most completely for copper–zinc superoxide dismutase SOD1. This oxidized form of this mutant protein has been detected in the spinal cord of ALS patients, where SOD1 mutations account for around 10% of total diagnoses, and variable SOD1 oxidation has also been implicated involved in sporadic forms of ALS. Recent studies conducted in sporadic ALS patients using mass spectrometry observed that SOD1 exhibits oxidation of three crucial residues, cysteine 146, histidine 71, and histidine 120, which are essential for protein stability, structural folding, and enzymatic activity (Martins & English, 2014).

Protein tyrosine nitration (PTN) due to the effects of nitric oxide (NO) is a common event in CNS tissues (Lee, Kim, Lee, & Kim, 2009; Schopfer, Baker, & Freeman, 2003). Accordingly, mass spectrometry has been used to identify an increased number of PTNs in the proteins flotillin-1 and α-tubulin in the brains of aging rodents (Dremina, Sharov, & Schoneich, 2005). The detrimental effects of protein nitration have been linked to diverse neurodegenerative diseases including AD, PD, HD, and ALS (Beal, 2002; Sacksteder et al., 2006; Sawada et al., 2007; Schopfer et al., 2003). In addition, increases in protein PTNs are strongly associated with the presentation of oxidative stress in neurons and glial cells (Ischiropoulos & Beckman, 2003). Furthermore, the presence of PTNs also seems to be linked with neuroinflammation (Glass, Saijo, Winner, Marchetto, & Gage, 2010) and NO-induced PTNs in the amino terminal of the tau protein have been reported to induce conformational changes and molecular modifications that promote tangle formation in the temporal regions of AD brain tissues (Reynolds, Berry, & Binder, 2005).

In the case of PD, the protein α-synuclein has been found to be nitrated at the tyrosine residues 39, 125, 133, and 136, leading to dramatically decreased solubility and protein aggregation (Takahashi, Yamashita, Nakamura, Nagano, & Nakamura, 2002). Additionally, SOD1 PTN induced by the enzymatic activity of the dismutase has also been reported in ALS, although nitration at tyrosine 108 does not appear to affect protein structure or function in this instance (Yeo, Kim, Kabir, Kang, & Kim, 2014).

5.2 Racemization and Isomerization

Racemization and isomerization are DPMs that occur at asparagine and glutamine residues and have been detected in isolated deposits in brain tissues from AD patients. Previous studies have shown that racemization of the aspartate residues in β-amyloid proteins can promote amyloidosis (Mori et al., 1994; Tomiyama et al., 1994). Several studies have now demonstrated that accumulation of isomerized residues in the amyloid protein of AD brains can promote fibrillation and aggregation, leading to increased protein toxicity (Roher, Lowenson, Clarke, Wolkow, et al., 1993; Roher, Lowenson, Clarke, Woods, et al., 1993). Some authors have since suggested that the study of amyloid DPMs should be expanded to encompass the analysis of succinimide, since this DPM is an intermediate product of aspartic acid dehydration and acts as precursor to isomerization and racemization (Orpiszewski & Benson, 1999; Orpiszewski, Schormann, Kluve-Beckerman, Liepnieks, & Benson, 2000).

5.3 Carbamylation

Protein carbamylation is not currently considered to be a major contributor to the neurodegenerative process, since this modification instead appears to exert neuroprotective effects in PD, AD and motor neuron diseases (Byts & Siren, 2009; Lapchak, 2008). In spite of these apparent beneficial effects, the molecular mechanism(s) by which protein carbamylation protects against neural decline remains unknown and understudied. With recent advances in mass spectrometry-based proteomic approaches, it should now be possible to uncover the function of carbamylation in neurodegenerative diseases, and to potentially harness these effects for the purposes of therapy. However, these promising beneficial effects of protein carbamylation in laboratory studies of neurodegeneration must be weighed against reports that the disease-promoting effects of harmful environmental factors such as smoking and uremia are also mediated by protein carbamylation (Wang et al., 2007).

5.4 Carbonylation

Previous studies have shown that induction of acute oxidative stress in the brain of *Macaca fuscata* leads to upregulation of carbonylated proteins, especially in the midbrain region (Oikawa et al., 2014). These findings are comparable to results obtained in PD brains in which an increase in protein carbonyls was detected in all brain regions (Alam et al., 1997). According to the conclusions drawn by Alam et al. (1997) in their study, the observed increase in carbonyl residues in the brain regions of PD patients cannot easily be attributed to the pathology of sporadic disease, since this could alternatively be explained by the pro-oxidant properties of the L-DOPA treatment that these patients receive. However, a subsequent proteomics study described the extent of protein carbonylation in AD brain tissues and reported that the proteins carbonic anhydrase II, syntaxin-binding protein I, heat-shock protein 70, and mitogen-activated protein kinase I were highly carbonylated in the brain regions first affected by AD (Sultana et al., 2010). In their report, the authors proposed that early carbonylation of these proteins could contribute to the early pathogenesis of AD. The importance of protein carbonylation in the pathology of neurodegenerative diseases is therefore clear, although it remains uncertain whether this modification is a consequence of the disease process or therapy given. Further studies will now be required to clarify this situation and to establish whether carbonylation is a feature of AD, the corresponding treatment, or even normal redox signaling processes in neurons (Suzuki, Carini, & Butterfield, 2010).

5.5 Advanced Glycation End-Products

AGEs have traditionally been regarded as a hallmark of acute oxidative stress during neurodegeneration, but as yet there have been no functional consequences identified for this DPM. It was recently proposed that AGEs may represent a source of mitogenic factors for neurons and could promote reentry into the cell cycle (Kuhla, Ludwig, Kuhla, Munch, & Vollmar, 2014; Schmidt, Kuhla, Bigl, Munch, & Arendt, 2007). Neurodegeneration is associated with expansion of glial cells, which could potentially cause proliferation of AGEs in the affected brain regions as it has been shown by Sasaki and colleagues. These authors found that astrocytes show positive signal for AGEs in brain regions devastated by the pathology of AD; furthermore, they and other authors suggest that a myriad of proteins may be modified by AGEs inside astrocytes during AD, and the details still remain unknown (Sasaki et al., 2001). Finally, we have to mention that the suppression of AGE formation by α-lipoic acid treatment has shown promising beneficial results *in vivo* during neurodegeneration (Kuhla et al., 2014).

6. PROTEIN DYSFUNCTIONS AND COGNITIVE IMPAIRMENT IN NEURODEGENERATIVE DISEASES

One of the most enigmatic phenomena in the pathology of neurodegenerative diseases is the appearance of clear and in some cases huge amounts of protein aggregates in neurons as well as in the extracellular space. Whether these aggregates are linked to the cognitive deficits that characterize dementia and/or whether they are involved in the long-term degeneration of affected neurons is still a controversial matter. Protein misfolding is a natural process that is not normally pathological, since around 30% of the total proteins synthesized in healthy cells are misfolded and must be degraded via normal mechanisms (Schubert et al., 2000). In addition, there is accumulating evidence that almost all proteins in mammalian cells can form amyloid-like aggregates under the appropriate conditions (Dobson, 2003; Fowler et al., 2006; Fowler, Koulov, Balch, & Kelly, 2007; Iconomidou et al., 2006; Mackay et al., 2001). Indeed, this hypothesis has been clearly established in yeast where inheritance of stable epigenetic traits is mediated by the transmission of cytoplasmic aggregated proteins (Serio & Lindquist, 2000). As we described above, the role of DPMs in the pathophysiological process of neurodegeneration remains elusive. Unbiased proteomic and systems biology studies of the neurodegenerative diseases are likely to generate valuable new data that can provide unforeseen insight into the molecular mechanism underlying these complex diseases. Recently, several studies have shown that sequestration of soluble toxic proteins in the cytoplasm to form insoluble aggregates can protect cells against cytotoxicity and reverses proteasome dysfunction (Bodner et al., 2006). Moreover, it has been shown that in HD, cells that exhibit this type of inclusion die later than do cells that lack the aggregates (Arrasate, Mitra, Schweitzer, Segal, & Finkbeiner, 2004; Schaffar et al., 2004). The harmful effects of protein aggregation may therefore depend on the extent of aggregation; when a cell is overloaded with oligomerized "pathological" protein, it is possible that this initiates an abnormal autophagic flux state that allows small aggregates to be assembled into big aggresomes that significantly impair normal cell function (Chen, Retzlaff, Roos, & Frydman, 2011; Swart et al., 2014). Current research on protein aggregation in neurodegenerative diseases focuses more on the study of oligomeric intermediate proteins and their toxic effects (Conway et al., 2000; Glabe, 2006; Swart et al., 2014). The influence of spontaneous DPMs on oligomeric intermediate proteins has been scarcely studied at all, although some reports have implied that nitration and oxidation may increase the aggregation capacity of α-synuclein in PD (Xiang et al., 2013). There are currently no conclusive data on whether spontaneous nonenzymatic DPMs to oligomeric intermediates can modulate cellular toxicity in neurodegenerative diseases.

The term oligomeric intermediates essentially defines proteins with toxic properties and high probability to be sequestered or form aggregates in neurons. Generally, the proteins studied and defined as oligomeric intermediates in almost all studies are cytoplasmic proteins (α-synuclein and SOD1), and little is known about the dysfunctions that occur following oligomerization of other important synaptic proteins, ion channel proteins, and mitochondrial components.

Patients suffering from many of the neurodegenerative diseases including AD, PD, frontotemporal-dementia-ALS, and HD manifest cognitive deficits that evolve into dementia. Cortical and subcortical dementias confer huge economic and social burdens on society, as well as severe consequences for the affected patients who may even lose their self-identity in the latter stages of disease (Hsiao, Kaiser, Fong, & Mendez, 2013). Degeneration of synapses is the first event in brains suffering from dementia, whereas the burden of senile plaques and amyloidosis in the brain does not correlate with dementia symptoms (Overk & Masliah, 2014). Accordingly, the extent of synapse loss correlates more strongly with the scores obtained in classical neuropsychological tests than do any other pathophysiological feature of disease (Dekosky & Scheff, 1990; Serrano-Pozo, Frosch, Masliah, & Hyman, 2011; Sze, Bi, Kleinschmidt-Demasters, Filley, & Martin, 2000; Sze et al., 1997). Given this apparent central role in neurodegenerative pathology, there is an urgent need to implement new studies based on unbiased approaches such as proteomics to discover the functional consequences of spontaneous DPMs on synaptic decay in dementia syndromes.

6.1 Deamidation of Ion Channel Proteins Na(+)K(+)-ATPase and Dementia

The multiple functions of the ion channel protein Na^+/K^+-ATPase include the maintenance of the differential membrane potential in neurons, which is an essential feature of signal transduction process (Mobasheri et al., 2000). The pump-action mechanism of this protein also appears to mediate the transport of nutrients, glucose, and other molecules along the axons and cytoplasmic membrane (Pavlov & Sokolov, 2000; Fig. 3).

Reduced expression levels of Na^+/K^+-ATPase have been reported in AD, PD, and HD in both animal models and human brain tissues (Cannon, 2004; Chauhan, Lee, & Siegel, 1997; Kumar & Kurup, 2002). Other studies have also analyzed and associated the levels of Na^+/K^+-ATPase with severe neuronal loss in rat brains by using pharmacological inactivation of the protein (Bignami & Palladini, 1966; Lees & Leong, 1994). Since Na(+)K(+)-ATPase is ubiquitously expressed in astrocytes and other brain cells, it was not possible until very recently to globally inhibit this protein and observe the effect on neuronal function (Harik, 1986; Renkawek, Renier, De Pont, Vogels, & Gabreels, 1992). However, the use of the eye neuronal system of fruit flies (*Drosophila melanogaster*) finally revealed that acute deficiency of this protein causes neuronal dysfunction, and that sustained-in-time deficit of Na(+)K(+)-ATPase triggers progressive neurodegeneration in this model (Luan, Reddig, & Li, 2014).

Although the link between dysregulation of Na(+)K(+)-ATPase and progressive neurodegeneration appears well established, dysfunctions in this protein had not been demonstrated in human dementia syndromes until 2014. Our recent iTRAQ quantitative proteomics study revealed

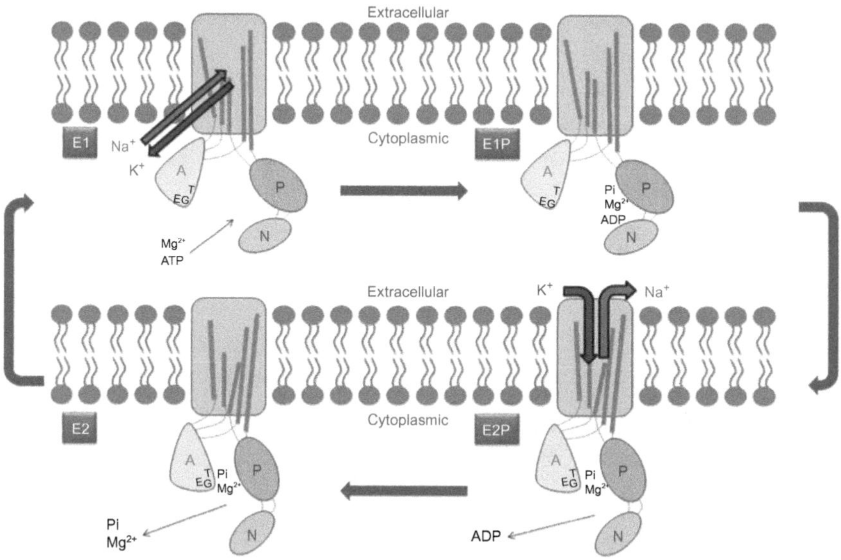

Figure 3 Simplified schematic of sodium/potassium ion transport in Na^+/K^+-ATPase. The α-subunit consists of four domains: N (nucleotide binding), A (actuator), P (phosphorylation), and M (membrane) domains. The major conformational changes during E1P to E2P transition are the 35° rotation of P domain, the rotation of N domain away from the P domain, and the rotation and translation of A domain that relocates the TGE sequence near to the catalytic site (Kuhlbrandt, 2004; Nyblom et al., 2013; Shinoda et al., 2009).

deamidation in the asparagine and glutamine residues of Na(+)K(+)-ATPase in the postmortem brain tissues of VaD patients (Adav et al., 2014). Deamidation of proteins added a negative charge at the affected residue and could lead to beta-isomerization. The impact of protein deamidation has been extensively studied and reported by Robinson and Robinson (2001a), who reported that deamidation of asparagine residues causes time-dependent changes in the charge profile and alters the three-dimensional conformation of several human proteins (Robinson & Robinson, 2001b; Robinson, Robinson, & Merrifield, 2001).

According to our analysis of VaD brain tissues, deamidation of Na(+) K(+)-ATPase subunits in the evolutionarily conserved regions of the protein alters intracellular electrostatic interactions and leads to conformational changes at the Mg^{2+}, Cu^{2+} binding site. Three of the deamidated residues identified in our study were located close to critical components of the Na^+/K^+-ATPase structure that would be expected to compromise the ion transport activity of the protein (Adav et al., 2014). It has been previously shown that mutation of Asn^{715} to Ala can decrease the affinity of Na^+/K^+-ATPase for Mg^{2+} ions (Pedersen, Jorgensen, & Jorgensen, 2000). Structural modeling experiments showed that the three deamidated residues present in VaD samples were located near to the catalytic site of this protein. In addition, modification of residues 210 and 220 has been proposed to cause defects in normal protein phosphorylation and dephosphorylation mechanisms, leading to altered ATP hydrolysis. The identification of these DPMs in the Na^+/K^+-ATPase ion channel was successfully uncovered by proteomic profiling, but since they have yet to be evaluated using alternative approaches, the roles played by these DPMs in VaD pathogenesis are currently unknown. Further application of unbiased proteomics approaches to the study of dementia syndromes is likely to provide unprecedented detail about the molecular pathology of neurodegenerative diseases (Fig. 4).

6.2 Deamidation of Synaptic Proteins and Dementia

Synaptic failure is the most common pathophysiological characteristic shared by both VaD and AD (Iadecola, 2013; Kalaria, 2000, 2002). Accordingly, loss of synapses has been reported as one of the most significant contributors to the cognitive impairments manifest in VaD and other neurodegenerative diseases. It has also been shown that following synaptic loss, the remaining synapses alter their shape as a potential compensatory mechanism (Scheff, 2001; Scheff, Dekosky, & Price, 1990; Scheff, Price, Schmitt, Dekosky, & Mufson, 2007; Sisková, Sanyal, Orban, O'connor, & Perry, 2010). However, the vast majority of studies performed in this field to date have been focused on the structural analysis of degenerating neuronal synapses using only imaging techniques. Consequently, the molecular profile associated with synaptic failure in VaD remains largely unknown. Our group pioneered the study of DPMs, dysfunctions, and deregulations in synaptic proteins in human VaD (Gallart-Palau et al., 2015). According to our quantitative findings using iTRAQ proteomics to profile human VaD postmortem brain tissues, synaptic immunoglobulins were among the most perturbed proteins in VaD temporal cortices, whereas synaptic proteins such as Bassoon and SNAP25 were substantially upregulated. In contrast, VaD brain tissues also displayed marked downregulation of the synaptic proteins AP2 and clathrin heavy chain 1, which are required for clathrin-mediated endocytosis.

When we extended our study to the analysis of synaptic protein deamidation in VaD temporal cortices, we observed that the protein synapsin1 displayed significant accumulation of deamidated asparagine and

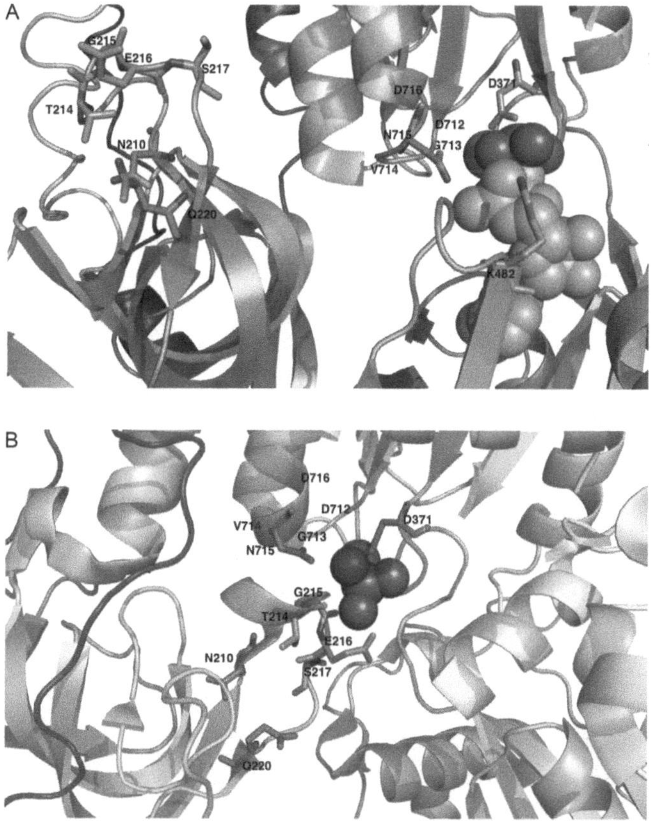

Figure 4 Structural models of Na^+/K^+-ATPase catalytic site in (A) E1P (Nyblom et al., 2013) and (B) E2P (Shinoda et al., 2009) conformations (PDB ID 4HQJ and 2ZXE, respectively). In both (A) and (B), color changes gradually from amino terminal (blue (dark gray in the print version)) to carboxyl terminal (red (gray in the print version)): A domain is in blue (dark gray in the print version) and cyan (light gray in the print version), P domain is in yellow (light gray in the print version), and N domain is in green (light gray in the print version). Notable deamidated residues, N210, D220, and N715, are in red (gray in the print version), whereas highly conserved sequences, ^{214}TGES, ^{712}DGVND, Asp^{371}, and Lys^{482}, are in gray. Magnesium ion is in magenta (dark gray in the print version), ADP is in orange (light gray in the print version), and phosphate analogue is in purple (dark gray in the print version). In E2P state (B), ^{712}TGES is located in close proximity to the catalytic site compared to E1P state (A). Structural models are visualized and prepared in The PyMOL Molecular Graphics System, Version 1.3 Schrödinger, LLC.

glutamine residues when compared with age-matched controls (Gallart-Palau et al., 2015). We next used the I-Tasser server to perform *ab initio* protein modeling (De Carvalho & De Mesquita, 2013; Roy, Kucukural, & Zhang, 2010; Zhang, 2008) which enabled us to construct models of both

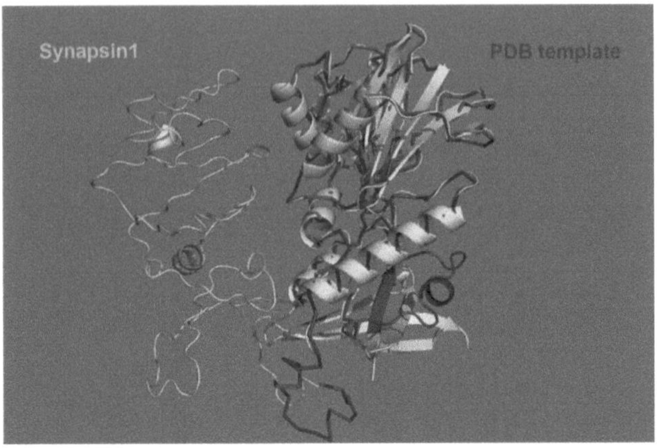

Figure 5 Structural modeling of the potential impact of deamidation sites on the conformation of synapsin1 protein. *Adapted from Gallart-Palau et al. (2015).*

wild-type synapsin1 and the deamidated form which was detected in VaD tissues. Next, we aligned the structure of the deamidated synapsin1 to the physical templates of wild-type synapsin1 (obtained by nuclear magnetic resonance using the software TM-align; Zhang & Skolnick, 2005). This then allowed to assess the likely impact of VaD-associated deamidations on the structure and function of the protein, and confirmed that the deamidation sites in synapsin1 were likely to induce pathological changes in protein conformation (Fig. 5).

The spontaneous protein deamidation events and other DPMs identified in dementia and neurodegeneration so far represent only the tip of the iceberg. Further studies using unbiased proteomics and systems biology approaches will now be needed to provide novel insight into the role of protein dysfunction in neurodegeneration, and should finally establish whether DPMs are merely the consequence of the ongoing neurodegenerative processes or are in fact essential mediators of these pathologies.

7. CONCLUDING REMARKS

The study of protein dysfunction in CNS tissues dates back more than 150 years, but progress in understanding the role of protein dysfunction in the pathology of neurodegenerative diseases has required substantial advancement in biochemical analytical methods and techniques. Mass spectrometry-based proteomics is an unbiased systems biological technique that enables the study of protein composition, characteristics, and interactions in both healthy tissues and pathological conditions. Application of mass spectrometry-based proteomics to the study of neurobiology has already facilitated numerous significant discoveries with the potential for therapeutic applications, as we have detailed in this chapter. However, it is the unforeseen insight that proteomics may provide in future years into the molecular basis of neurodegenerative diseases that could deliver radically improved patient outcomes in the clinic.

ACKNOWLEDGEMENTS

We thank the Singapore National Medical Research Council (NMRC/CBRG/0004/2012) and the Singapore National Research Foundation (NRF2011 NRF-CRP 001-109) for financial support.

REFERENCES

Abdelmegeed, M. A., Jang, S., Banerjee, A., Hardwick, J. P., & Song, B. J. (2013). Robust protein nitration contributes to acetaminophen-induced mitochondrial dysfunction and acute liver injury. *Free Radical Biology & Medicine, 60*, 211–222. http://dx.doi.org/10.1016/j.freeradbiomed.2013.02.018.

Adams, R. N., Murrill, E., McCreery, R., Blank, L., & Karolczak, M. (1972). 6-Hydroxydopamine, a new oxidation mechanism. *European Journal of Pharmacology, 17*, 287–292.

Adav, S. S., Qian, J., Ang, Y. L., Kalaria, R. N., Lai, M. K., Chen, C. P., et al. (2014). iTRAQ quantitative clinical proteomics revealed role of Na(+)K(+)-ATPase and its correlation with deamidation in vascular dementia. *Journal of Proteome Research, 13*, 4635–4646. http://dx.doi.org/10.1021/pr500754j.

Aksenov, M. Y., Aksenova, M. V., Butterfield, D. A., Geddes, J. W., & Markesbery, W. R. (2001). Protein oxidation in the brain in Alzheimer's disease. *Neuroscience, 103*, 373–383.

Alam, Z. I., Daniel, S. E., Lees, A. J., Marsden, D. C., Jenner, P., & Halliwell, B. (1997). A generalised increase in protein carbonyls in the brain in Parkinson's but not incidental Lewy body disease. *Journal of Neurochemistry, 69*, 1326–1329.

Arrasate, M., Mitra, S., Schweitzer, E. S., Segal, M. R., & Finkbeiner, S. (2004). Inclusion body formation reduces levels of mutant huntingtin and the risk of neuronal death. *Nature, 431*, 805–810. http://dx.doi.org/10.1038/nature02998.

Bayes, A., & Grant, S. G. (2009). Neuroproteomics: Understanding the molecular organization and complexity of the brain. *Nature Reviews. Neuroscience, 10*, 635–646. http://dx.doi.org/10.1038/nrn2701.

Beal, M. F. (2002). Oxidatively modified proteins in aging and disease. *Free Radical Biology & Medicine, 32*, 797–803.

Berlett, B. S., & Stadtman, E. R. (1997). Protein oxidation in aging, disease, and oxidative stress. *The Journal of Biological Chemistry, 272*, 20313–20316.

Bignami, A., & Palladini, G. (1966). Experimentally produced cerebral status spongiosus and continuous pseudorhythmic electroencephalographic discharges with a membrane-ATPase inhibitor in the rat. *Nature, 209*, 413–414.

Bodner, R. A., Outeiro, T. F., Altmann, S., Maxwell, M. M., Cho, S. H., Hyman, B. T., et al. (2006). Pharmacological promotion of inclusion formation: A therapeutic approach for Huntington's and Parkinson's diseases. *Proceedings of the National Academy of Sciences of the United States of America, 103*, 4246–4251. http://dx.doi.org/10.1073/pnas.0511256103.

Boyd-Kimball, D., Sultana, R., Poon, H. F., Lynn, B. C., Casamenti, F., Pepeu, G., et al. (2005). Proteomic identification of proteins specifically oxidized by intracerebral injection of amyloid beta-peptide (1-42) into rat brain: Implications for Alzheimer's disease. *Neuroscience, 132*, 313–324. http://dx.doi.org/10.1016/j.neuroscience.2004.12.022.

Brownlee, M., Cerami, A., & Vlassara, H. (1988). Advanced glycosylation end products in tissue and the biochemical basis of diabetic complications. *The New England Journal of Medicine, 318*, 1315–1321. http://dx.doi.org/10.1056/nejm198805193182007.

Bucala, R., Makita, Z., Koschinsky, T., Cerami, A., & Vlassara, H. (1993). Lipid advanced glycosylation: Pathway for lipid oxidation in vivo. *Proceedings of the National Academy of Sciences of the United States of America, 90*, 6434–6438.

Butterfield, D. A. (2003). Amyloid beta-peptide [1-42]-associated free radical-induced oxidative stress and neurodegeneration in Alzheimer's disease brain: Mechanisms and consequences. *Current Medicinal Chemistry, 10*, 2651–2659.

Butterfield, D. A., Drake, J., Pocernich, C., & Castegna, A. (2001). Evidence of oxidative damage in Alzheimer's disease brain: Central role for amyloid beta-peptide. *Trends in Molecular Medicine, 7*, 548–554.

Butterfield, D. A., & Lauderback, C. M. (2002). Lipid peroxidation and protein oxidation in Alzheimer's disease brain: Potential causes and consequences involving amyloid beta-peptide-associated free radical oxidative stress. *Free Radical Biology & Medicine, 32*, 1050–1060.

Byts, N., & Siren, A. L. (2009). Erythropoietin: A multimodal neuroprotective agent. *Experimental & Translational Stroke Medicine, 1*, 4. http://dx.doi.org/10.1186/2040-7378-1-4.

Cain, J. A., Solis, N., & Cordwell, S. J. (2014). Beyond gene expression: The impact of protein post-translational modifications in bacteria. *Journal of Proteomics, 97*, 265–286. http://dx.doi.org/10.1016/j.jprot.2013.08.012.

Cannon, S. C. (2004). Paying the price at the pump: Dystonia from mutations in a Na^+/K^+-ATPase. *Neuron, 43*, 153–154. http://dx.doi.org/10.1016/j.neuron.2004.07.002.

Chait, B. T. (2006). Chemistry. Mass spectrometry: Bottom-up or top-down? *Science, 314*(5796), 65–66.

Chauhan, N. B., Lee, J. M., & Siegel, G. J. (1997). Na, K-ATPase mRNA levels and plaque load in Alzheimer's disease. *Journal of Molecular Neuroscience, 9*, 151–166. http://dx.doi.org/10.1007/bf02800498.

Chen, B., Retzlaff, M., Roos, T., & Frydman, J. (2011). Cellular strategies of protein quality control. *Cold Spring Harbor Perspectives in Biology, 3*, a004374. http://dx.doi.org/10.1101/cshperspect.a004374.

Choudhary, J., & Grant, S. G. (2004). Proteomics in postgenomic neuroscience: The end of the beginning. *Nature Neuroscience, 7*, 440–445. http://dx.doi.org/10.1038/nn1240.

Clarke, S. (1987). Propensity for spontaneous succinimide formation from aspartyl and asparaginyl residues in cellular proteins. *International Journal of Peptide and Protein Research, 30*, 808–821.

Cohen, A. S. (1967). Amyloidosis. *The New England Journal of Medicine, 277*, 522–530. http://dx.doi.org/10.1056/NEJM196709072771006.

Cohen, A. S., & Calkins, E. (1959). Electron microscopic observations on a fibrous component in amyloid of diverse origins. *Nature, 183*, 1202–1203.

Conway, K. A., Lee, S. J., Rochet, J. C., Ding, T. T., Williamson, R. E., & Lansbury, P. T., Jr. (2000). Acceleration of oligomerization, not fibrillization, is a shared property of both alpha-synuclein mutations linked to early-onset Parkinson's disease: Implications for pathogenesis and therapy. *Proceedings of the National Academy of Sciences of the United States of America, 97*, 571–576.

Craft, G. E., Chen, A., & Nairn, A. C. (2013). Recent advances in quantitative neuroproteomics. *Methods*, *61*, 186–218. http://dx.doi.org/10.1016/j.ymeth.2013.04.008.

Dang, X., Scotcher, J., Wu, S., Chu, R. K., Tolic, N., Ntai, I., et al. (2014). The first pilot project of the consortium for top-down proteomics: A status report. *Proteomics*, *14*, 1130–1140. http://dx.doi.org/10.1002/pmic.201300438.

D'angelo, S., Ingrosso, D., Migliardi, V., Sorrentino, A., Donnarumma, G., Baroni, A., et al. (2005). Hydroxytyrosol, a natural antioxidant from olive oil, prevents protein damage induced by long-wave ultraviolet radiation in melanoma cells. *Free Radical Biology & Medicine*, *38*, 908–919. http://dx.doi.org/10.1016/j.freeradbiomed.2004.12.015.

Datta, A., Akatsu, H., Heese, K., & Sze, S. K. (2013). Quantitative clinical proteomic study of autopsied human infarcted brain specimens to elucidate the deregulated pathways in ischemic stroke pathology. *Journal of Proteomics*, *91*, 556–568. http://dx.doi.org/10.1016/j.jprot.2013.08.017.

Datta, A., Chen, C. P., & Sze, S. K. (2014). Discovery of prognostic biomarker candidates of lacunar infarction by quantitative proteomics of microvesicles enriched plasma. *PLoS One*, *9*, e94663. http://dx.doi.org/10.1371/journal.pone.0094663.

Datta, A., Jingru, Q., Khor, T. H., Teo, M. T., Heese, K., & Sze, S. K. (2011). Quantitative neuroproteomics of an in vivo rodent model of focal cerebral ischemia/reperfusion injury reveals a temporal regulation of novel pathophysiological molecular markers. *Journal of Proteome Research*, *10*, 5199–5213. http://dx.doi.org/10.1021/pr200673y.

Datta, A., Park, J. E., Li, X., Zhang, H., Ho, Z. S., Heese, K., et al. (2010). Phenotyping of an in vitro model of ischemic penumbra by iTRAQ-based shotgun quantitative proteomics. *Journal of Proteome Research*, *9*, 472–484. http://dx.doi.org/10.1021/pr900829h.

Datta, A., Qian, J., Chong, R., Kalaria, R. N., Francis, P., Lai, M. K., et al. (2014). Novel pathophysiological markers are revealed by iTRAQ-based quantitative clinical proteomics approach in vascular dementia. *Journal of Proteomics*, *99*, 54–67. http://dx.doi.org/10.1016/j.jprot.2014.01.011.

De Carvalho, M. D. C., & De Mesquita, J. F. (2013). Structural modeling and in silico analysis of human superoxide dismutase 2. *PLoS One*, *8*, e65558. http://dx.doi.org/10.1371/journal.pone.0065558.

Dekosky, S. T., & Scheff, S. W. (1990). Synapse loss in frontal cortex biopsies in Alzheimer's disease: Correlation with cognitive severity. *Annals of Neurology*, *27*, 457–464. http://dx.doi.org/10.1002/ana.410270502.

Desrosiers, R. R., & Fanélus, I. (2011). Damaged proteins bearing L-isoaspartyl residues and aging: A dynamic equilibrium between generation of isomerized forms and repair by PIMT. *Current Aging Science*, *4*, 8–18.

Dobson, C. M. (2003). Protein folding and misfolding. *Nature*, *426*, 884–890. http://dx.doi.org/10.1038/nature02261.

Dremina, E. S., Sharov, V. S., & Schoneich, C. (2005). Protein tyrosine nitration in rat brain is associated with raft proteins, flotillin-1 and alpha-tubulin: Effect of biological aging. *Journal of Neurochemistry*, *93*, 1262–1271. http://dx.doi.org/10.1111/j.1471-4159.2005.03115.x.

Dukan, S., Farewell, A., Ballesteros, M., Taddei, F., Radman, M., & Nystrom, T. (2000). Protein oxidation in response to increased transcriptional or translational errors. *Proceedings of the National Academy of Sciences of the United States of America*, *97*, 5746–5749. http://dx.doi.org/10.1073/pnas.100422497.

Dutta, B., Yan, R., Lim, S. K., Tam, J. P., & Sze, S. K. (2014). Quantitative profiling of chromatome dynamics reveals a novel role for HP1BP3 in hypoxia-induced oncogenesis. *Molecular & Cellular Proteomics*, *13*, 3236–3249. http://dx.doi.org/10.1074/mcp.M114.038232.

Erill, S., Calvo, R., & Carlos, R. (1980). Plasma protein carbamylation and decreased acidic drug protein binding in uremia. *Clinical Pharmacology and Therapeutics*, *27*, 612–618.

Fabian, H., Szendrei, G. I., Mantsch, H. H., Greenberg, B. D., & Otvos, L., Jr. (1994). Synthetic post-translationally modified human A beta peptide exhibits a markedly increased tendency to form beta-pleated sheets in vitro. *European Journal of Biochemistry*, *221*, 959–964.

Fischer, E. H., Graves, D. J., Crittenden, E. R., & Krebs, E. G. (1959). Structure of the site phosphorylated in the phosphorylase b to a reaction. *The Journal of Biological Chemistry*, *234*, 1698–1704.

Fluckiger, R., Harmon, W., Meier, W., Loo, S., & Gabbay, K. H. (1981). Hemoglobin carbamylation in uremia. *The New England Journal of Medicine*, *304*, 823–827. http://dx.doi.org/10.1056/nejm198104023041406.

Fowler, D. M., Koulov, A. V., Alory-Jost, C., Marks, M. S., Balch, W. E., & Kelly, J. W. (2006). Functional amyloid formation within mammalian tissue. *PLoS Biology*, *4*, e6. http://dx.doi.org/10.1371/journal.pbio.0040006.

Fowler, D. M., Koulov, A. V., Balch, W. E., & Kelly, J. W. (2007). Functional amyloid—From bacteria to humans. *Trends in Biochemical Sciences*, *32*, 217–224. http://dx.doi.org/10.1016/j.tibs.2007.03.003.

Gallart-Palau, X., Serra, A., Qian, J., Chen, C. P., Kalaria, R. N., & Sze, S. K. (2015). Temporal lobe proteins implicated in synaptic failure exhibit differential expression and deamidation in vascular dementia. *Neurochemistry International*, *80*, 87–98. http://dx.doi.org/10.1016/j.neuint.2014.12.002.

Gillery, P., & Jaisson, S. (2013). Usefulness of non-enzymatic post-translational modification derived products (PTMDPs) as biomarkers of chronic diseases. *Journal of Proteomics*, *92*, 228–238. http://dx.doi.org/10.1016/j.jprot.2013.02.015.

Gillery, P., & Jaisson, S. (2014). Post-translational modification derived products (PTMDPs): Toxins in chronic diseases? *Clinical Chemistry and Laboratory Medicine*, *52*, 33–38. http://dx.doi.org/10.1515/cclm-2012-0880.

Glabe, C. G. (2006). Common mechanisms of amyloid oligomer pathogenesis in degenerative disease. *Neurobiology of Aging*, *27*, 570–575. http://dx.doi.org/10.1016/j.neurobiolaging.2005.04.017.

Glass, C. K., Saijo, K., Winner, B., Marchetto, M. C., & Gage, F. H. (2010). Mechanisms underlying inflammation in neurodegeneration. *Cell*, *140*, 918–934. http://dx.doi.org/10.1016/j.cell.2010.02.016.

Gupta, R., & Srivastava, O. P. (2004). Deamidation affects structural and functional properties of human alphaA-crystallin and its oligomerization with alphaB-crystallin. *The Journal of Biological Chemistry*, *279*, 44258–44269. http://dx.doi.org/10.1074/jbc.M405648200.

Halim, A., Carlsson, M. C., Madsen, C. B., Brand, S., Moller, S. R., Olsen, C. E., et al. (2015). Glycoproteomic analysis of seven major allergenic proteins reveals novel post-translational modifications. *Molecular & Cellular Proteomics*, *14*, 191–204. http://dx.doi.org/10.1074/mcp.M114.042614.

Halliwell, B. (2006). Oxidative stress and neurodegeneration: Where are we now? *Journal of Neurochemistry*, *97*, 1634–1658. http://dx.doi.org/10.1111/j.1471-4159.2006.03907.x.

Han, X., Jin, M., Breuker, K., & McLafferty, F. W. (2006). Extending top-down mass spectrometry to proteins with masses greater than 200 kilodaltons. *Science*, *314*(5796), 109–112.

Hao, P., Ren, Y., Alpert, A. J., & Sze, S. K. (2011). Detection, evaluation and minimization of nonenzymatic deamidation in proteomic sample preparation. *Molecular & Cellular Proteomics: MCP*, *10*. http://dx.doi.org/10.1074/mcp.O111.009381. O111.009381.

Hao, P., & Sze, S. K. (2014). Proteomic analysis of protein deamidation. *Current Protocols in Protein Science*, *78*, 24.5.1–24.5.14. http://dx.doi.org/10.1002/0471140864.ps2405s78.

Harik, S. I. (1986). Blood–brain barrier sodium/potassium pump: Modulation by central noradrenergic innervation. *Proceedings of the National Academy of Sciences of the United States of America*, *83*, 4067–4070.

Hipkiss, A. R. (2006). Accumulation of altered proteins and ageing: Causes and effects. *Experimental Gerontology*, *41*, 464–473. http://dx.doi.org/10.1016/j.exger.2006.03.004.

Hsiao, J. J., Kaiser, N., Fong, S. S., & Mendez, M. F. (2013). Suicidal behavior and loss of the future self in semantic dementia. *Cognitive and Behavioral Neurology*, *26*, 85–92. http://dx.doi.org/10.1097/WNN.0b013e31829c671d.

Iadecola, C. (2013). The pathobiology of vascular dementia. *Neuron*, *80*, 844–866. http://dx.doi.org/10.1016/j.neuron.2013.10.008.

Iconomidou, V. A., Chryssikos, G. D., Gionis, V., Galanis, A. S., Cordopatis, P., Hoenger, A., et al. (2006). Amyloid fibril formation propensity is inherent into the hexapeptide tandemly repeating sequence of the central domain of silkmoth chorion proteins of the A-family. *Journal of Structural Biology*, *156*, 480–488. http://dx.doi.org/10.1016/j.jsb.2006.08.011.

Ischiropoulos, H., & Beckman, J. S. (2003). Oxidative stress and nitration in neurodegeneration: Cause, effect, or association? *The Journal of Clinical Investigation*, *111*, 163–169. http://dx.doi.org/10.1172/jci17638.

Jenner, P., & Olanow, C. W. (1998). Understanding cell death in Parkinson's disease. *Annals of Neurology*, *44*, S72–S84.

Kalaria, R. N. (2000). The role of cerebral ischemia in Alzheimer's disease. *Neurobiology of Aging*, *21*, 321–330.

Kalaria, R. N. (2002). Small vessel disease and Alzheimer's dementia: Pathological considerations. *Cerebrovascular Diseases*, *13*(Suppl. 2), 48–52. http://dx.doi.org/10.1159/000049150.

Kruger, R., Kuhn, W., Muller, T., Woitalla, D., Graeber, M., Kosel, S., et al. (1998). Ala30Pro mutation in the gene encoding alpha-synuclein in Parkinson's disease. *Nature Genetics*, *18*, 106–108. http://dx.doi.org/10.1038/ng0298-106.

Kuhla, A., Ludwig, S. C., Kuhla, B., Munch, G., & Vollmar, B. (2014). Advanced glycation end products are mitogenic signals and trigger cell cycle reentry of neurons in Alzheimer's disease brain. *Neurobiology of Aging*, *36*, 753–761. http://dx.doi.org/10.1016/j.neurobiolaging.2014.09.025.

Kuhlbrandt, W. (2004). Biology, structure and mechanism of P-type ATPases. *Nature Reviews. Molecular Cell Biology*, *5*(4), 282–295.

Kumar, A. R., & Kurup, P. A. (2002). Endogenous sodium-potassium ATPase inhibition related biochemical cascade in trisomy 21 and Huntington's disease: Neural regulation of genomic function. *Neurology India*, *50*, 174–180.

Lapchak, P. A. (2008). Carbamylated erythropoietin to treat neuronal injury: New development strategies. *Expert Opinion on Investigational Drugs*, *17*, 1175–1186. http://dx.doi.org/10.1517/13543784.17.8.1175.

Lavoie, M. J., Ostaszewski, B. L., Weihofen, A., Schlossmacher, M. G., & Selkoe, D. J. (2005). Dopamine covalently modifies and functionally inactivates parkin. *Nature Medicine*, *11*, 1214–1221. http://dx.doi.org/10.1038/nm1314.

Lee, J. R., Kim, J. K., Lee, S. J., & Kim, K. P. (2009). Role of protein tyrosine nitration in neurodegenerative diseases and atherosclerosis. *Archives of Pharmacal Research*, *32*, 1109–1118. http://dx.doi.org/10.1007/s12272-009-1802-0.

Lees, G. J., & Leong, W. (1994). Brain lesions induced by specific and non-specific inhibitors of sodium-potassium ATPase. *Brain Research*, *649*, 225–233.

Lim, K.-L., Chua, D. S. K., Palau, X. G., & Yao, T.-P. (2014). Chapter 11—PARK2 induces autophagy removal of impaired mitochondria via ubiquitination. In M. A. Hayat (Ed.), *Autophagy: Cancer, other pathologies, inflammation, immunity, infection, and aging* (pp. 175–188). Amsterdam: Academic Press.

Lowenson, J., & Clarke, S. (1988). Does the chemical instability of aspartyl and asparaginyl residues in proteins contribute to erythrocyte aging? The role of protein carboxyl methylation reactions. *Blood Cells*, *14*, 103–118.

Luan, Z., Reddig, K., & Li, H. S. (2014). Loss of Na(+)/K(+)-ATPase in Drosophila photoreceptors leads to blindness and age-dependent neurodegeneration. *Experimental Neurology, 261*, 791–801. http://dx.doi.org/10.1016/j.expneurol.2014.08.025.

Lubec, G., Krapfenbauer, K., & Fountoulakis, M. (2003). Proteomics in brain research: Potentials and limitations. *Progress in Neurobiology, 69*, 193–211.

Mackay, J. P., Matthews, J. M., Winefield, R. D., Mackay, L. G., Haverkamp, R. G., & Templeton, M. D. (2001). The hydrophobin EAS is largely unstructured in solution and functions by forming amyloid-like structures. *Structure, 9*, 83–91.

Mann, M., & Jensen, O. N. (2003). Proteomic analysis of post-translational modifications. *Nature Biotechnology, 21*, 255–261. http://dx.doi.org/10.1038/nbt0303-255.

Martins, D., & English, A. M. (2014). SOD1 oxidation and formation of soluble aggregates in yeast: Relevance to sporadic ALS development. *Redox Biology, 2*, 632–639. http://dx.doi.org/10.1016/j.redox.2014.03.005.

Masliah, E., Alford, M., Deteresa, R., Mallory, M., & Hansen, L. (1996). Deficient glutamate transport is associated with neurodegeneration in Alzheimer's disease. *Annals of Neurology, 40*, 759–766. http://dx.doi.org/10.1002/ana.410400512.

Merkley, E. D., Metz, T. O., Smith, R. D., Baynes, J. W., & Frizzell, N. (2014). The succinated proteome. *Mass Spectrometry Reviews, 33*, 98–109. http://dx.doi.org/10.1002/mas.21382.

Mobasheri, A., Avila, J., Cozar-Castellano, I., Brownleader, M. D., Trevan, M., Francis, M. J., et al. (2000). Na^+, K^+-ATPase isozyme diversity; comparative biochemistry and physiological implications of novel functional interactions. *Bioscience Reports, 20*, 51–91.

Mommen, G. P., Frese, C. K., Meiring, H. D., Van Gaans-Van Den Brink, J., De Jong, A. P., Van Els, C. A., et al. (2014). Expanding the detectable HLA peptide repertoire using electron-transfer/higher-energy collision dissociation (EThcD). *Proceedings of the National Academy of Sciences of the United States of America, 111*, 4507–4512. http://dx.doi.org/10.1073/pnas.1321458111.

Moradian, A., Kalli, A., Sweredoski, M. J., & Hess, S. (2014). The top-down, middle-down, and bottom-up mass spectrometry approaches for characterization of histone variants and their post-translational modifications. *Proteomics, 14*, 489–497. http://dx.doi.org/10.1002/pmic.201300256.

Mori, H., Ishii, K., Tomiyama, T., Furiya, Y., Sahara, N., Asano, S., et al. (1994). Racemization: Its biological significance on neuropathogenesis of Alzheimer's disease. *The Tohoku Journal of Experimental Medicine, 174*, 251–262.

Nyblom, M., Poulsen, H., Gourdon, P., Reinhard, L., Andersson, M., Lindahl, E., et al. (2013). Crystal Structure of Na^+, K^+-ATPase in the Na^+-Bound State. *Science, 342*(6154), 123–127.

Oikawa, S., Kobayashi, H., Kitamura, Y., Zhu, H., Obata, K., Minabe, Y., et al. (2014). Proteomic analysis of carbonylated proteins in the monkey substantia nigra after ischemia-reperfusion. *Free Radical Research, 48*, 694–705. http://dx.doi.org/10.3109/10715762.2014.901509.

Olsen, J. V., & Mann, M. (2013). Status of large-scale analysis of post-translational modifications by mass spectrometry. *Molecular & Cellular Proteomics, 12*, 3444–3452. http://dx.doi.org/10.1074/mcp.O113.034181.

Olsen, J. V., Ong, S. E., & Mann, M. (2004). Trypsin cleaves exclusively C-terminal to arginine and lysine residues. *Molecular & Cellular Proteomics, 3*, 608–614. http://dx.doi.org/10.1074/mcp.T400003-MCP200.

Ong, S. E., & Mann, M. (2005). Mass spectrometry-based proteomics turns quantitative. *Nature Chemical Biology, 1*, 252–262. http://dx.doi.org/10.1038/nchembio736.

Orpiszewski, J., & Benson, M. D. (1999). Induction of beta-sheet structure in amyloidogenic peptides by neutralization of aspartate: A model for amyloid nucleation. *Journal of Molecular Biology, 289*, 413–428. http://dx.doi.org/10.1006/jmbi.1999.2768.

Orpiszewski, J., Schormann, N., Kluve-Beckerman, B., Liepnieks, J. J., & Benson, M. D. (2000). Protein aging hypothesis of Alzheimer disease. *FASEB Journal: Official Publication of the Federation of American Societies for Experimental Biology*, *14*, 1255–1263.

Ott, C., Jacobs, K., Haucke, E., Navarrete Santos, A., Grune, T., & Simm, A. (2014). Role of advanced glycation end products in cellular signaling. *Redox Biology*, *2*, 411–429. http://dx.doi.org/10.1016/j.redox.2013.12.016.

Overk, C. R., & Masliah, E. (2014). Pathogenesis of synaptic degeneration in Alzheimer's disease and Lewy body disease. *Biochemical Pharmacology*, *88*, 508–516. http://dx.doi.org/10.1016/j.bcp.2014.01.015.

Pavlov, K. V., & Sokolov, V. S. (2000). Electrogenic ion transport by Na^{+}, K^{+}-ATPase. *Membrane & Cell Biology*, *13*, 745–788.

Pedersen, P. A., Jorgensen, J. R., & Jorgensen, P. L. (2000). Importance of conserved alpha - subunit segment 709GDGVND for Mg^{2+} binding, phosphorylation, and energy transduction in Na, K-ATPase. *The Journal of Biological Chemistry*, *275*, 37588–37595. http://dx.doi.org/10.1074/jbc.M005610200.

Powell, J. T., Vine, N., & Crossman, M. (1992). On the accumulation of D-aspartate in elastin and other proteins of the ageing aorta. *Atherosclerosis*, *97*, 201–208.

Renkawek, K., Renier, W. O., De Pont, J. J., Vogels, O. J., & Gabreels, F. J. (1992). Neonatal status convulsivus, spongiform encephalopathy, and low activity of Na+/K(+)-ATPase in the brain. *Epilepsia*, *33*, 58–64.

Reynolds, M. R., Berry, R. W., & Binder, L. I. (2005). Site-specific nitration and oxidative dityrosine bridging of the tau protein by peroxynitrite: Implications for Alzheimer's disease. *Biochemistry*, *44*, 1690–1700. http://dx.doi.org/10.1021/bi047982v.

Ritz, S., Turzynski, A., Schutz, H. W., Hollmann, A., & Rochholz, G. (1996). Identification of osteocalcin as a permanent aging constituent of the bone matrix: Basis for an accurate age at death determination. *Forensic Science International*, 77, 13–26.

Ritz-Timme, S., & Collins, M. J. (2002). Racemization of aspartic acid in human proteins. *Ageing Research Reviews*, *1*, 43–59.

Robinson, N. E., & Robinson, A. B. (2001a). Deamidation of human proteins. *Proceedings of the National Academy of Sciences of the United States of America*, *98*, 12409–12413. http://dx.doi.org/10.1073/pnas.221463198.

Robinson, N. E., & Robinson, A. B. (2001b). Prediction of protein deamidation rates from primary and three-dimensional structure. *Proceedings of the National Academy of Sciences of the United States of America*, *98*, 4367–4372. http://dx.doi.org/10.1073/pnas.071066498.

Robinson, N. E., Robinson, A. B., & Merrifield, R. B. (2001). Mass spectrometric evaluation of synthetic peptides as primary structure models for peptide and protein deamidation. *The Journal of Peptide Research*, *57*, 483–493.

Roher, A. E., Lowenson, J. D., Clarke, S., Wolkow, C., Wang, R., Cotter, R. J., et al. (1993). Structural alterations in the peptide backbone of beta-amyloid core protein may account for its deposition and stability in Alzheimer's disease. *The Journal of Biological Chemistry*, *268*, 3072–3083.

Roher, A. E., Lowenson, J. D., Clarke, S., Woods, A. S., Cotter, R. J., Gowing, E., et al. (1993). beta-Amyloid-(1-42) is a major component of cerebrovascular amyloid deposits: Implications for the pathology of Alzheimer disease. *Proceedings of the National Academy of Sciences of the United States of America*, *90*, 10836–10840.

Ross, P. L., Huang, Y. N., Marchese, J. N., Williamson, B., Parker, K., Hattan, S., et al. (2004). Multiplexed protein quantitation in Saccharomyces cerevisiae using amine-reactive isobaric tagging reagents. *Molecular & Cellular Proteomics*, *3*, 1154–1169. http://dx.doi.org/10.1074/mcp.M400129-MCP200.

Roy, A., Kucukural, A., & Zhang, Y. (2010). I-TASSER: A unified platform for automated protein structure and function prediction. *Nature Protocols*, *5*, 725–738. http://dx.doi.org/10.1038/nprot.2010.5.

Ruderman, N. B., Williamson, J. R., & Brownlee, M. (1992). Glucose and diabetic vascular disease. *FASEB Journal: Official Publication of the Federation of American Societies for Experimental Biology*, *6*, 2905–2914.
Sacksteder, C. A., Qian, W. J., Knyushko, T. V., Wang, H., Chin, M. H., Lacan, G., et al. (2006). Endogenously nitrated proteins in mouse brain: Links to neurodegenerative disease. *Biochemistry*, *45*, 8009–8022. http://dx.doi.org/10.1021/bi060474w.
Sakurai, T., Sugioka, K., & Nakano, M. (1990). O2- generation and lipid peroxidation during the oxidation of a glycated polypeptide, glycated polylysine, in the presence of iron-ADP. *Biochimica et Biophysica Acta*, *1043*, 27–33.
Sasaki, N., Toki, S., Chowei, H., Saito, T., Nakano, N., Hayashi, Y., et al. (2001). Immunohistochemical distribution of the receptor for advanced glycation end products in neurons and astrocytes in Alzheimer's disease. *Brain Research*, *888*, 256–262.
Sawada, H., Ishiguro, H., Nishii, K., Yamada, K., Tsuchida, K., Takahashi, H., et al. (2007). Characterization of neuron-specific huntingtin aggregates in human huntingtin knock-in mice. *Neuroscience Research*, *57*, 559–573. http://dx.doi.org/10.1016/j.neures.2007.01.002.
Schaffar, G., Breuer, P., Boteva, R., Behrends, C., Tzvetkov, N., Strippel, N., et al. (2004). Cellular toxicity of polyglutamine expansion proteins: Mechanism of transcription factor deactivation. *Molecular Cell*, *15*, 95–105. http://dx.doi.org/10.1016/j.molcel.2004.06.029.
Scheff, S. (2001). Quantitative assessment of possible age-related change in synaptic numbers in the human frontal cortex. *Neurobiology of Aging*, *22*, 355–365. http://dx.doi.org/10.1016/S0197-4580(01)00222-6.
Scheff, S. W., Dekosky, S. T., & Price, D. A. (1990). Quantitative assessment of cortical synaptic density in Alzheimer's disease. *Neurobiology of Aging*, *11*, 29–37. http://dx.doi.org/10.1016/0197-4580(90)90059-9.
Scheff, S. W., Price, D. A., Schmitt, F. A., Dekosky, S. T., & Mufson, E. J. (2007). Synaptic alterations in CA1 in mild Alzheimer disease and mild cognitive impairment. *Neurology*, *68*, 1501–1508. http://dx.doi.org/10.1212/01.wnl.0000260698.46517.8f.
Schmidt, A., Kuhla, B., Bigl, K., Munch, G., & Arendt, T. (2007). Cell cycle related signaling in Neuro2a cells proceeds via the receptor for advanced glycation end products. *Journal of Neural Transmission*, *114*, 1413–1424. http://dx.doi.org/10.1007/s00702-007-0770-0.
Schopfer, F. J., Baker, P. R., & Freeman, B. A. (2003). NO-dependent protein nitration: A cell signaling event or an oxidative inflammatory response? *Trends in Biochemical Sciences*, *28*, 646–654. http://dx.doi.org/10.1016/j.tibs.2003.10.006.
Schubert, U., Anton, L. C., Gibbs, J., Norbury, C. C., Yewdell, J. W., & Bennink, J. R. (2000). Rapid degradation of a large fraction of newly synthesized proteins by proteasomes. *Nature*, *404*, 770–774. http://dx.doi.org/10.1038/35008096.
Schuessler, H., & Schilling, K. (1984). Oxygen effect in the radiolysis of proteins. Part 2. Bovine serum albumin. *International Journal of Radiation Biology and Related Studies in Physics, Chemistry, and Medicine*, *45*, 267–281.
Selkoe, D. J. (2001). Alzheimer's disease results from the cerebral accumulation and cytotoxicity of amyloid beta-protein. *Journal of Alzheimer's Disease*, *3*, 75–80.
Serio, T. R., & Lindquist, S. L. (2000). Protein-only inheritance in yeast: Something to get [PSI+]-ched about. *Trends in Cell Biology*, *10*, 98–105.
Serrano-Pozo, A., Frosch, M. P., Masliah, E., & Hyman, B. T. (2011). Neuropathological alterations in Alzheimer disease. *Cold Spring Harbor Perspectives in Medicine*, *1*, a006189. http://dx.doi.org/10.1101/cshperspect.a006189.
Shao, C. H., Capek, H. L., Patel, K. P., Wang, M., Tang, K., Desouza, C., et al. (2011). Carbonylation contributes to SERCA2a activity loss and diastolic dysfunction in a rat model of type 1 diabetes. *Diabetes*, *60*, 947–959. http://dx.doi.org/10.2337/db10-1145.
Shinoda, T., Ogawa, H., Cornelius, F., & Toyoshima, C. (2009). Crystal structure of the sodium-potassium pump at 2.4 A resolution. *Nature*, *459*(7245), 446–450.

Sipe, J. D., & Cohen, A. S. (2000). Review: History of the amyloid fibril. *Journal of Structural Biology*, *130*, 88–98. http://dx.doi.org/10.1006/jsbi.2000.4221.

Sisková, Z., Sanyal, N. K., Orban, A., O'connor, V., & Perry, V. H. (2010). Reactive hypertrophy of synaptic varicosities within the hippocampus of prion-infected mice. *Biochemical Society Transactions*, *38*, 471–475. http://dx.doi.org/10.1042/BST0380471.

Smith, L. M., & Kelleher, N. L. (2013). Proteoform: a single term describing protein complexity. *Nature Methods*, *10*(3), 186–187.

Soskic, V., Groebe, K., & Schrattenholz, A. (2008). Nonenzymatic posttranslational protein modifications in ageing. *Experimental Gerontology*, *43*, 247–257. http://dx.doi.org/10.1016/j.exger.2007.12.001.

Souza, J. M., Choi, I., Chen, Q., Weisse, M., Daikhin, E., Yudkoff, M., et al. (2000). Proteolytic degradation of tyrosine nitrated proteins. *Archives of Biochemistry and Biophysics*, *380*, 360–366. http://dx.doi.org/10.1006/abbi.2000.1940.

Sultana, R., Perluigi, M., Newman, S. F., Pierce, W. M., Cini, C., Coccia, R., et al. (2010). Redox proteomic analysis of carbonylated brain proteins in mild cognitive impairment and early Alzheimer's disease. *Antioxidants & Redox Signaling*, *12*, 327–336. http://dx.doi.org/10.1089/ars.2009.2810.

Suzuki, Y. J., Carini, M., & Butterfield, D. A. (2010). Protein carbonylation. *Antioxidants & Redox Signaling*, *12*, 323–325. http://dx.doi.org/10.1089/ars.2009.2887.

Swart, C., Haylett, W., Kinnear, C., Johnson, G., Bardien, S., & Loos, B. (2014). Neurodegenerative disorders: Dysregulation of a carefully maintained balance? *Experimental Gerontology*, *58*, 279–291. http://dx.doi.org/10.1016/j.exger.2014.09.003.

Sze, C. I., Bi, H., Kleinschmidt-Demasters, B. K., Filley, C. M., & Martin, L. J. (2000). Selective regional loss of exocytotic presynaptic vesicle proteins in Alzheimer's disease brains. *Journal of the Neurological Sciences*, *175*, 81–90.

Sze, S. K., Ge, Y., Oh, H., & McLafferty, F. W. (2002). Top-down mass spectrometry of a 29-kDa protein for characterization of any posttranslational modification to within one residue. *Proceedings of the National Academy of Sciences of the United States of America*, *99*(4), 1774–1779.

Sze, C. I., Troncoso, J. C., Kawas, C., Mouton, P., Price, D. L., & Martin, L. J. (1997). Loss of the presynaptic vesicle protein synaptophysin in hippocampus correlates with cognitive decline in Alzheimer disease. *Journal of Neuropathology and Experimental Neurology*, *56*, 933–944.

Takahashi, T., Yamashita, H., Nakamura, T., Nagano, Y., & Nakamura, S. (2002). Tyrosine 125 of alpha-synuclein plays a critical role for dimerization following nitrative stress. *Brain Research*, *938*, 73–80.

Terry, R. D., Masliah, E., Salmon, D. P., Butters, N., Deteresa, R., Hill, R., et al. (1991). Physical basis of cognitive alterations in Alzheimer's disease: Synapse loss is the major correlate of cognitive impairment. *Annals of Neurology*, *30*, 572–580. http://dx.doi.org/10.1002/ana.410300410.

Tomiyama, T., Asano, S., Furiya, Y., Shirasawa, T., Endo, N., & Mori, H. (1994). Racemization of Asp23 residue affects the aggregation properties of Alzheimer amyloid beta protein analogues. *The Journal of Biological Chemistry*, *269*, 10205–10208.

Tran, J. C., Zamdborg, L., Ahlf, D. R., Lee, J. E., Catherman, A. D., Durbin, K. R., et al. (2011). Mapping intact protein isoforms in discovery mode using top-down proteomics. *Nature*, *480*, 254–258. http://dx.doi.org/10.1038/nature10575.

Van Kasteren, S. I., Kramer, H. B., Jensen, H. H., Campbell, S. J., Kirkpatrick, J., Oldham, N. J., et al. (2007). Expanding the diversity of chemical protein modification allows post-translational mimicry. *Nature*, *446*, 1105–1109. http://dx.doi.org/10.1038/nature05757.

Varadarajan, S., Yatin, S., Aksenova, M., & Butterfield, D. A. (2000). Review: Alzheimer's amyloid beta-peptide-associated free radical oxidative stress and neurotoxicity. *Journal of Structural Biology*, *130*, 184–208. http://dx.doi.org/10.1006/jsbi.2000.4274.

Victorino, V. J., Mencalha, A. L., & Panis, C. (2014). Post-translational modifications disclose a dual role for redox stress in cardiovascular pathophysiology. *Life Sciences*, *129*, 42–47. http://dx.doi.org/10.1016/j.lfs.2014.11.008.

Wang, Z., Nicholls, S. J., Rodriguez, E. R., Kummu, O., Horkko, S., Barnard, J., et al. (2007). Protein carbamylation links inflammation, smoking, uremia and atherogenesis. *Nature Medicine*, *13*, 1176–1184. http://dx.doi.org/10.1038/nm1637.

Xiang, W., Schlachetzki, J. C., Helling, S., Bussmann, J. C., Berlinghof, M., Schaffer, T. E., et al. (2013). Oxidative stress-induced posttranslational modifications of alpha-synuclein: Specific modification of alpha-synuclein by 4-hydroxy-2-nonenal increases dopaminergic toxicity. *Molecular and Cellular Neurosciences*, *54*, 71–83. http://dx.doi.org/10.1016/j.mcn.2013.01.004.

Yan, S. D., Chen, X., Schmidt, A. M., Brett, J., Godman, G., Zou, Y. S., et al. (1994). Glycated tau protein in Alzheimer disease: A mechanism for induction of oxidant stress. *Proceedings of the National Academy of Sciences of the United States of America*, *91*, 7787–7791.

Yan, S. D., Schmidt, A. M., Anderson, G. M., Zhang, J., Brett, J., Zou, Y. S., et al. (1994). Enhanced cellular oxidant stress by the interaction of advanced glycation end products with their receptors/binding proteins. *The Journal of Biological Chemistry*, *269*, 9889–9897.

Yeo, W. S., Kim, Y. J., Kabir, M. H., Kang, J. W., & Kim, K. P. (2014). Mass spectrometric analysis of protein tyrosine nitration in aging and neurodegenerative diseases. *Mass Spectrometry Reviews*, *34*, 166–183. http://dx.doi.org/10.1002/mas.21429.

Zhang, Y. (2008). I-TASSER server for protein 3D structure prediction. *BMC Bioinformatics*, *9*, 40. http://dx.doi.org/10.1186/1471-2105-9-40.

Zhang, Y., & Skolnick, J. (2005). TM-align: A protein structure alignment algorithm based on the TM-score. *Nucleic Acids Research*, *33*, 2302–2309. http://dx.doi.org/10.1093/nar/gki524.

CHAPTER FIVE

Comparative Proteomics for the Evaluation of Protein Expression and Modifications in Neurodegenerative Diseases

Antonio Conti, Massimo Alessio[1]
Proteome Biochemistry, Division of Genetics and Cell Biology, IRCCS San Raffaele Scientific Institute, Milano, Italy
[1]Corresponding author: e-mail address: m.alessio@hsr.it

Contents

1. Introduction 118
2. Essential Neuroproteomics 119
   2.1 Differential Quantitative Proteomics 119
   2.2 Sample Tissues 122
   2.3 Immune Proteomics 123
   2.4 Biocomputational Analysis 126
   2.5 Analysis of PTMs 127
3. Conclusions 144
References 145

Abstract

Together with hypothesis-driven approaches, high-throughput differential proteomic analysis performed primarily not only in human cerebrospinal fluid and serum but also on protein content of other tissues (blood cells, muscles, peripheral nerves, etc.) has been used in the last years to investigate neurodegenerative diseases. Even if the goal for these analyses was mainly the discovery of neurodegenerative disorders biomarkers, the characterization of specific posttranslational modifications (PTMs) and the differential protein expression resulted in being very informative to better define the pathological mechanisms. In this chapter are presented and discussed the positive aspects and challenges of the outcomes of some of our investigations on neurological and neurodegenerative disease, in order to highlight the important role of protein PTMs studies in proteomics-based approaches.

International Review of Neurobiology, Volume 121
ISSN 0074-7742
http://dx.doi.org/10.1016/bs.irn.2015.05.004

1. INTRODUCTION

The term "proteome" refers to the entire complement of proteins, including the different posttranslational modifications (PTMs), expressed by cells or homogeneous tissues in a certain time. Therefore, the proteome is a dynamic concept that varies with time and events occurring to cells or organisms, whereas the term "proteomics" encompasses the study of a specific proteome, including information on protein abundances, their variations, and modifications, together with the analysis of their interaction with partners and networks, in order to understand cellular processes. In addition, "clinical proteomics" is a branch of proteomics that involves the application of proteomic technologies on clinical specimens such as blood, cerebrospinal fluid (CSF), sera, and tissue samples.

The proteome shows an increased degree of complexity compared to the genome; in fact, one gene can encode more than one differently functional protein (even up to 1000 for a single gene); therefore, since the human genome contains about 21,000 protein-encoding genes, the total number of proteins in human cells is estimated to be in a range of 250,000 to 1 million. In addition, proteins are dynamic, undergoing synthesis and degradation, binding to the cell membranes, or interacting with other proteins to form complexes, while the genome, on the other hand, is relatively static.

The functional dynamics of proteins is primarily regulated by the PTMs that occur as consequences of many different conditions such as the individual genetic identity, the gender, the age, and the different kinds of tissues, or as consequences of different environmental conditions like temperature, stress factors, energy availability (food intake), and, not least, pathological conditions. Thus, protein expression levels and PTMs can vary considerably not only in different individuals in the same environmental conditions but also within the same individual at different ages or states of health. Another element of complexity in proteomics studies is the fact that proteins exist in an extremely wide range of concentrations in the body, from millimolar (10^{-3}) to attomolar (10^{-18}), facing the investigators with instrumental and technical limitations. For example, the concentration of the protein albumin in sera and CSF is much greater than the majority of all other proteins, making it extremely difficult to detect the low-abundance proteins.

The proteomics approach has been applied to neurodegenerative diseases mainly to identify biomarkers for early or differential diagnosis, and for prognosis in order to address prompt clinical treatments (Zhang, Keene,

Pan, Montine, & Montine, 2008). Certainly, neurodegenerative disorders would benefit from the identification of such biomarkers, even to see how the patients respond to a specific treatment or therapy. However, the vast majority of the new biomarkers proposed in the scientific literature prove to be weak. The difficulty to translate this data into useful tools for the clinical neurologist is mostly due to the fact that proteomics approaches are useful for the selection of "candidate biomarkers," but requires standardized validation studies on appropriate case series. These validation studies are performed using nonproteomics technologies (e.g., ELISA, RIA, etc.) and usually are out of the interests of the basic science groups that discovered the candidate biomarkers. To allow the encouraging results to become useful clinical tools, investments are necessary by the academy, together with the industry in order to generate laboratory dedicated to the validation analysis. These dedicated laboratory should deal with practical problems like the fact that preliminary proteomics investigations can be done with very different methodological approaches, which in turn may result in very different conclusions; second, clinical validation of biomarkers needs large cohorts of patients that are difficult to put together in single centers; further, they should take into consideration that in proteomics studies preanalytical errors, which account for most of the analytical errors (Plebani, Sciacovelli, Aita, Padoan, & Chiozza, 2014) and affect several aspects (sample selection in multicentric studies, time and type of processing, storage conditions, etc.), are rarely considered. Therefore, in these "validation laboratories," an implementation of international guidelines, like in the case of CSF (Bossuyt et al., 2003; Teunissen et al., 2009), is mandatory to strengthen studies and focus investments on solid results.

In our opinion, more fruitful and interesting appear to be the results of proteomics analysis in which the variation of protein expression levels and PTMs is used to gain insight into and understand the pathological mechanisms, reflecting changes in certain pathways during the disease.

2. ESSENTIAL NEUROPROTEOMICS

2.1 Differential Quantitative Proteomics

The classic proteomics approach used to compare two or more experimental conditions or group of samples is the differential expression analysis; in general, the investigated disease is compared with other similar diseases for differential diagnosis purposes, or compared with healthy control and/or experimental conditions generated *ex vivo* in order to discover pathological pathways and disease mechanism. Nowadays, the quantitative differential proteomics, which measure the relative and/or absolute amount of proteins or specific PTMs, is based on two methodological approaches that differ for the gel-based or liquid chromatography (LC)-based step of proteins/peptides fractionation prior to the mass spectrometry (MS) analysis and protein identification (for a detailed methodological review, see Craft, Chen, & Nairn, 2013; Giron, Dayon, Turck, Hoogland, & Sanchez, 2011; van Gool & Hendrickson, 2012). The gel-based methodologies, mainly two-dimensional electrophoresis (2DE), are still powerful and largely used, and they rely on intact proteins separation, protein quantification by gel-image analysis, digestion of relevant spots with trypsin, and then tandem MS analysis for protein identification performed either by MALDI-TOF/TOF (matrix-assisted laser desorption ionization-time of flight) or by ESI-MS/MS (electron spray ionization-MS), whereas the liquid chromatography (LC)-based approach is characterized by a protein digestion that precedes the peptides separation in liquid-phase chromatography and a direct ionization by ESI into tandem MS for protein identifications. Both approaches have advantages and disadvantages; in particular, the 2DE gel-based analysis is limited in the number and type of proteins that can be resolved, but conversely is able to reveal the presence of protein fragments and isoforms undetectable if peptides are analyzed as for LC–MS/MS; on the contrary, the MS approaches coupled with LC are able to resolve, without type-specific limitation, large number of proteins (Huang et al., 2007; Yuan & Desiderio, 2005a; Zhang et al., 2008). Usually, the relative quantitation of protein expression is reported as a fold change and can be obtained, within the two approaches, either by label-based or by label-free methods: typically 2DE gel staining versus fluorescence-labeled protein 2DE for gel-based approach, and label-free quantitative MS versus isotope-coded affinity tags for LC–MS approach (Andreev et al., 2012; Craft et al., 2013; Giron et al., 2011; Huang et al., 2007; van Gool & Hendrickson, 2012; Yuan & Desiderio, 2005a; Zhang et al., 2008). Absolute quantitation can be determined by addition in the samples of internal standards; in particular, the addition of stable isotopic standard is suitable in the LC–MS approach with the so-called iTRAQ (isobaric tag for relative and absolute quantitation) (reviewed in Craft et al., 2013).

Human brain homogenate has been used to investigate proteins in Alzheimer's disease (AD) patients in a number of studies that used different extraction protocols or analyzed different regions of the brain (Schonberger, Edgar, Kydd, Faull, & Cooper, 2001; Shiozaki et al., 2004). In these studies, several proteins variations have been identified; however, few of them resulted to be coherent in the different studies (e.g., the increased level of GAPDH and synaptotagmin I in AD). Other differential proteomics studies have been performed with the aim to characterize the protein components of different aggregates present in neurodegenerative diseases, for example, amyloid plaques and neurofibrillary tangles from AD patients and Lewy bodies from patients with dementia with Lewy bodies; most of these studies were performed using material enrichment by laser capture microdissection from human brain samples (Leverenz et al., 2007; Liao et al., 2004; Wakabayashi, Tanji, Mori, & Takahashi, 2007; Wang et al., 2005). A large bulk of studies have been done using the proteomics approaches on CSF in particular in AD (see Liu, Qing, & Deng, 2014 for review); depending on the methodology used, from 5 to more than 400 different proteins have been found altered in the CSF of AD patients compared to controls, including differential expression of protein isoforms when 2DE was used (Castano, Roher, Esh, Kokjohn, & Beach, 2006; Davidsson et al., 2002; Hu et al., 2007; Hu, Malone, Fagan, Townsend, & Holtzman, 2005; Lista, Zetterberg, Dubois, Blennow, & Hampel, 2014; Zhang, Goodlett, Peskind, et al., 2005; Zhang, Goodlett, Quinn, et al., 2005). Proteomics analysis has also been performed on the CSF from other neurodegenerative diseases like multiple sclerosis and amyotrophic lateral sclerosis (ALS) (Conti et al., 2008; Lourenco, Baldeiras, Graos, & Duarte, 2011; Stoop, Coulier, et al., 2010; Stoop, Singh, et al., 2010; see Craft et al., 2013 for a review). Since some neurodegenerative diseases are associated with human aging, of interest were also the studies that try to define the proteome changes in CSF associated with human aging (Zhang, Goodlett, Peskind, et al., 2005; Zhang, Goodlett, Quinn, et al., 2005).

Although it seems that an enormous number of candidate proteins have been discovered, only very few of them have been validated as putative biomarkers and the relevance of these finding remains to be defined, underling the need of a rigorous validation process.

In our opinion, in a nonhypothesis-driven shotgun proteomics approach, the 2DE methodology is more suitable for the identification of PTMs specifically associated with the disease conditions that can be further studied as to obtain indications on the pathological mechanisms (Rogowska-Wrzesinska, Le Bihan, Thaysen-Andersen, & Roepstorff, 2013).

2.2 Sample Tissues

Several kinds of tissues have been used for the human neurodegenerative diseases' proteomics analysis. For obvious reasons, brain proteins analysis has been performed exclusively from samples of patient autopsy both on "fresh" materials and on frozen or formalin-fixed specimens (Schonberger et al., 2001; Shiozaki et al., 2004). With this kind of samples, it is critical the time that occurs between the patients death and the collection and processing of the specimens; in fact, most of the brain protein postmortem modifications and degradation start as soon as death takes place, and after few hours, they dramatically affect the proteomics profile of the sample, making, in most of the case, the analysis useless. More reliable are the studies performed on protein aggregates characterizing the neurodegenerative conditions, like Lewy bodies, amyloid plaques, and neurofibrillary tangles (Cumming & Schubert, 2005; Leverenz et al., 2007; Liao et al., 2004; Wakabayashi et al., 2007; Wang et al., 2005) because these are more stable structures. Proteomics analysis of brain has been more frequently used when animal models of neurodegenerative diseases are investigated (Gotz & Ittner, 2008), while some studies have been performed on cellular models, principally cell lines as the dopaminergic neuroblastoma SH-SY5Y (Alberio et al., 2010; Fasano, Alberio, Colapinto, Mila, & Lopiano, 2008).

Proteomics analyses of specimens from human brain have been principally used for pathogenesis studies, while the standard tissue used for proteomics biomarker discovery in neurodegenerative diseases is the CSF (Blennow, 2005; D'Ascenzo, Relkin, & Lee, 2005; Davidsson & Sjogren, 2005). Indeed, it directly drains the target organ and may reflect the metabolic changes induced by the pathological conditions in the brain. However, as for brain tissue, some limitations may occur and, due to the biological and chemical features of the CSF, the introduction of a rigorous standardized experimental pipeline is mandatory to improve the quality of the results (Zhang, 2007). This includes specific sample preprocessing steps in order to deal with the high salt concentration, the relative low protein concentration, and the wide dynamic range (several order of magnitude) of some specific proteins concentration (Yuan & Desiderio, 2005a). The sample preprocessing and protein prefractionation can be achieved by standard biochemical methods that exploit different biochemical features of the protein like hydrophobicity, size, and charge (Waller, Shores, & Knapp, 2008; Yuan & Desiderio, 2005a, 2005b; Zhang, Goodlett, Peskind, et al., 2005; Zhang, Goodlett, Quinn, et al., 2005). For CSF proteomics analysis, other critical points that might dramatically affect the outcome of the analyses must be taken into consideration; these are the risk of contamination by blood proteins that may occur during sample collection, the interindividual variation of proteins concentrations, and the age-related changes in protein relative abundance (Perrin et al., 2013; Schutzer et al., 2010; Stoop, Coulier, et al., 2010; Stoop, Singh, et al., 2010; Zhang, Goodlett, Peskind, et al., 2005; Zhang, Goodlett, Quinn, et al., 2005).

Due to the privileged interaction of CSF with brain and its almost routinely availability (at least for some disease in which CSF is collected for differential diagnosis purposes), proteomics analyses of serum from neurodegenerative disease patients have rarely been conducted. On the contrary, some studies have been performed on blood mononuclear cells from Parkinson's disease (PD) patients (Mila et al., 2009). A promising compartment worth investigating in neurodegenerative diseases is the "secretome" that refers to the protein released by cells along the classical secretory pathway together with the protein content of the extracellular membrane vesicles/exosomes. It would be, for example, interesting to analyze the secretome of the choroid plexus epithelia cells that contribute to the proteins composition of CSF (Thouvenot et al., 2006). Finally, other tissues have been used according to specific features of the studied disease like, for example, the muscle specimens, recently used for investigating the ALS both in human (Conti et al., 2014) and in mice models (Capitanio et al., 2012).

2.3 Immune Proteomics

A role of neuroinflammation in contributing to the progression of neurodegenerative diseases like AD, PD, and ALS has been clearly reported (Glass, Saijo, Winner, Marchetto, & Gage, 2010); nowadays, together with the neuroinflammatory elements, a contribution to some disorders of an autoimmune component has also been hypothesized (El-Fawal, 2014). The hypothesis is that, in environmental toxic-induced neurodegeneration, alterations in intracellular structural proteins organization and proteolysis occur and that protein fragments are released into CSF and blood where they are processed by antigen-presenting cells in the lymphoid tissue to induce humoral autoimmune responses (El-Fawal, 2014). Defining these pathological effects mediated by the humoral immune response might provide accessible information for predicting and diagnosing neurological outcome depending on the neurotoxic effect.

An alternative proteomics approach that must be taken into consideration for neurodegenerative diseases with autoimmune component is the serological proteome analysis (SERPA) (Klade et al., 2001; Seliger & Kellner, 2002). SERPA approach exploits the presence of autoantibodies in the biological fluids of patients (serum, CSF, etc.) to screen the proteome of patient's cells/tissues resolved by 2DE, in order to identify putative autoantigen/s (Fig. 1). This kind of approach has been mainly used with success in cancer immunoproteomics, being the titer of specific autoantibodies

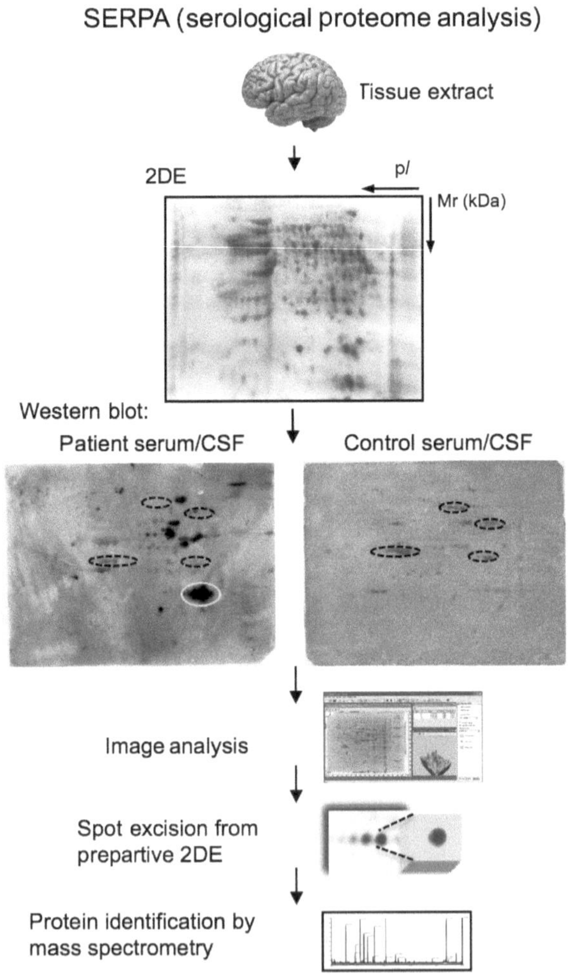

Figure 1 Schematic representation of the workflow for candidate antigens' identification by SERPA that exploits the presence of autoantibodies directed against brain proteins in the sera or cerebrospinal fluid of neurodegenerative disease patients.

considered itself a putative biomarker (Caron, Choquet-Kastylevsky, & Joubert-Caron, 2007). Furthermore, the immunogenicity of specific protein isoforms and the recognition of different autoantigen as a function of disease progression have been documented in cancer, highlighting the changes of the immune response during the disease evolvement (De Monte et al., 2008). After the candidate antigen/s identification by SERPA, it is possible to prepare the screening of large cohorts of patient sera to establish the reactivity against purified or recombinant relevant autoantigens using more suitable methodologies.

The presence in the CSF of autoantibodies and the identification of the immunogenic antigens have been investigated with SERPA also in some neurological disease, including multiple sclerosis, in which several antigenic candidates have been identified (Almeras et al., 2004; Gini et al., 2008; Menon et al., 2011). An early inflammatory phase has been postulated in several movement disorders, such as PD, and in the hyperkinetic movement disorders (HMDs) characterized by involuntary movements (Sanger et al., 2010). In HMDs, neuron dysfunction in the basal ganglia plays a pathogenic role, and autoantibodies directed against basal ganglia antigens have already been reported in the literature (Church et al., 2002; Dale, Heyman, Giovannoni, & Church, 2005). In order to ascertain the identity of a target antigen that is recognized by antibasal ganglia autoantibodies found in the serum from an HMD patient, we applied a SERPA approach testing by Western blot the sera reactivity against the mouse basal ganglia proteome resolved by 2DE. A unique specific reactivity against a spot of about 40 kDa yielded the identification by MS of the fructose-bisphosphate aldolase A enzyme as putative antigen. The specificity of this reactivity, as biomarker of autoimmunity in HMD, has been validated by ELISA using the purified aldolase A. Results indicated that anti-aldolase A antibodies were associated mainly to HMD (41%) and PD (13%) patients and were undetectable in patients with other inflammatory and noninflammatory central nervous system (CNS) diseases (Privitera et al., 2013). Aldolases are glycolytic enzymes, and their role in energy metabolism has been studied in neurodegenerative disorders characterized by oxidative stress. Two isoforms of aldolase, namely A and C, are expressed in the human brain with different distributions (Buono, D'Armiento, Terzi, Alfieri, & Salvatore, 2001). Interestingly, previous proteomics studies reported the occurrence of serum anti-aldolase C antibodies in PD, particularly in patients with atypical features (van de Warrenburg et al., 2008), while anti-aldolase A antibodies were found in patients with heterogeneous disorders, such as AD and multiple sclerosis (Lovato et al., 2008; Mor, Izak, & Cohen, 2005). The possible

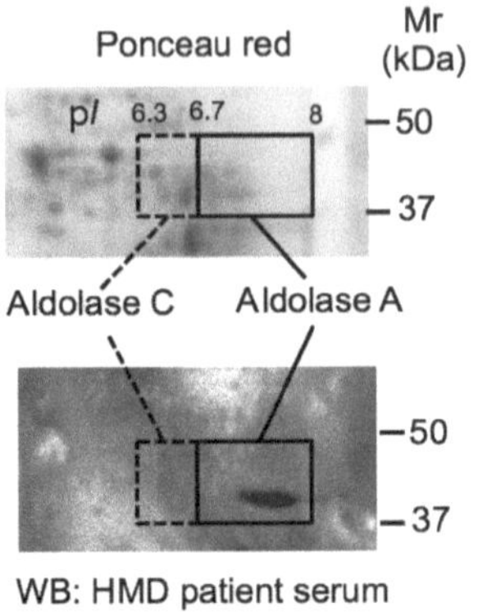

Figure 2 Aldolase isoform detection. Mouse brain basal ganglia proteins resolved by 2DE are transferred on to nitrocellulose membrane and visualized by Ponceau red staining. Western blot performed on the same material with sera from HMD patients specifically recognizes a protein spot (arrow) in the region of the gel corresponding to the theoretical p*I* of aldolase A (solid-line boxes), and the reactivity is not detectable in the region of the gel corresponding to the theoretical p*I* of aldolase C (dashed-line boxes).

cross-reactivity of the autoantibodies with the aldolase C isoform has been ruled out by the lack of reactivity against the recombinant aldolase C and by the 2DE electrophoretic features of the spot recognized by patients sera, the C isoform being more acidic than the A isoform (a range of 6.3–6.8 vs. 7.84–8.4, respectively) (Fig. 2). Since functional alterations in glycolytic pathways are known to occur in movement disorders (e.g., in the cerebral cortex in PD, even in the preclinical phase; Ferrer, Martinez, Blanco, Dalfo, & Carmona, 2011), we suggested that anti-aldolase A antibodies might have a role in these disorders. However, at the moment, it is not clear whether autoantibodies are an epiphenomenon that results from sustained damage and inflammation of the target tissues, or whether autoimmunity is one of the primary events of the disease with antibodies playing a direct pathogenic role. Nevertheless, autoantibody anti-aldolase A may represent a useful biomarker of autoimmunity in patient of HMD and patients with atypical movement disorders.

2.4 Biocomputational Analysis

A critical aspect for the proteomics approaches, but in general for all kinds of -omics approach, is the enormous amount of data generated by the experimental technologies, which in turn represents a challenge that needs complex statistical and bioinformatics analyses for the identification of the relevant features. Moreover, the general idea is that, for multifactorial diseases, like most of the neurodegenerative diseases, the ideal unique biomarker does not exist. Therefore, it may be necessary to find a pattern or a functional pathway signature, possibly of different molecular nature (proteins, metabolites, etc.), that would be more useful both for biomarker definition and for pathological mechanisms discovery. In addition to multivariate statistical analysis, that can be done using commercial or free available software, new approaches for unsupervised data mining such as neural networks and machine learning are increasingly used and/or in house implemented as computational pipelines (Bradley, 2012; Cannistraci, Alanis-Lobato, & Ravasi, 2013b; Cannistraci, Montevecchi, & Alessio, 2009; Cannistraci, Ravasi, Montevecchi, Ideker, & Alessio, 2010; Olivieri et al., 2011; Smolinska, Blanchet, Buydens, & Wijmenga, 2012). Linear dimensionality reduction approaches like principal component analysis are largely and successfully used for visualization and discrimination of patterns hidden in multidimensional proteomics dataset; however, they may present limitations in particular when samples are nonlinearly related, as often occurs in 2DE datasets. In particular, it must be taken into consideration the need to develop robust algorithms useful to discriminate sample groups also for data set with restricted number of elements (e.g., tens of patients) having large number of features (e.g., thousands of proteins) as frequently occurs for human specimens proteomics dataset (Cannistraci, Alanis-Lobato, & Ravasi, 2013a; Ravasi et al., 2010). We developed and used a nonlinear unsupervised machine learning for dimensionality reduction called minimum curvilinear embedding that, on the basis of the entire CSF protein profile, allowed us to discriminate the patients with peripheral neuropathies with pain from patients without pain (including healthy subjects and motor neuron disease patients) (Cannistraci et al., 2010).

2.5 Analysis of PTMs

In addition to the analysis of differential protein expression levels, and many times as an extension of the full comparative analysis, proteomics investigations of specific PTMs promoted by the pathological conditions have been performed. These studies resulted in some interesting observations that can be used both for understanding pathological mechanisms and as a putative source of biomarkers. The analysis of PTMs usually combines 2DE protein separation with a plethora of classical biochemical analysis that exploit specific enzymes or chemical reagents, or alternatively combines 2DE with ad hoc *ex vivo* experimental conditions able to exacerbate the conditions that are able to foster the PTM under investigation.

Accurate discrimination between closely related protein isoforms, which may differ by as little as a single amino acid or single-residue modification, is necessary for precise protein identification. The high resolving power of 2DE allows separation of protein isoforms in which modifications add or remove an electric charge, resulting in small changes in the net electric charge of proteins (Rogowska-Wrzesinska et al., 2013). A less sensitive method is the detection of small changes in molecular mass that can be further characterized by MS. Therefore, 2DE can provide indication of some PTMs, and among these, the most common are glycosylation, acetylation, phosphorylation, carbonylation, and deamidation.

In some neurodegenerative diseases, accumulation of protein pathologically modified has been reported; in particular, oxidative modifications have been investigated in several diseases characterized by the presence of neuroinflammation or/and oxidative stress conditions. For example, several studies have been conducted on immunodetection of protein carbonyls in 2DE followed by MS for the identification of oxidized proteins in the brain extracts of AD patients (Choi et al., 2004). In the same disorder, the detection of nitrotyrosine residues allowed identification of α-enolase, triosephosphate isomerase, and neuropolypeptide h3 as target for this type of pathological modification in the cerebral cortex of patients (Butterfield & Sultana, 2008a, 2008b; Castegna et al., 2002). These data were in part confirmed by nitrotyrosine analysis performed on brain autopsy samples from mild cognitive impairment (MCI) patients, a disorder condition that frequently progresses toward AD and can be somehow considered an early stage of AD (Butterfield et al., 2007; Sultana et al., 2007). Protein nitration has also been investigated in the CSF during aging and in the CSF of AD patients and was found to be increased compared to control subjects (Tohgi et al., 1999). These finding suggest an activation of tyrosine nitration, or an increase in nitrated tyrosine-containing proteins, and/or reduced degradation may be involved in brain aging and may play a role in AD pathology.

Other specific protein reactive groups can be the target of the oxidative modifications, like the thiol-group, that can undergo several modifications (S-cysteinylation, S-glutathionylation, and S-nitrosylation). These residues have been found modified in the transthyretin protein in the CSF of AD and MCI patients (Nakamura et al., 2013; Poulsen, Bahl, Simonsen,

Hasselbalch, & Heegaard, 2014). Moreover, the most common PTM used for the reversible regulation of protein function, namely the protein phosphorylation, has also been investigated by 2DE and Western blot in normal CSF (Yuan & Desiderio, 2003).

To illustrate how proteomic studies of PTMs are useful for making progress to understanding the mechanism of neurological diseases and discovering potential therapeutic targets, we will use examples of our own research conducted on central or peripheral neurological diseases, including also note of cautions on possible critical technical aspects.

2.5.1 Differential Analysis of CSF's Protein Profile in Traumatic Brain Injury Patients

Classical 2DE differential proteomics analysis of the CSF from severe traumatic brain injury (TBI) patients was carried out with the aim to find trauma-associated markers that can eventually correlate with prognosis. Like other CNS pathologies, TBI causes change in the composition of CSF as the results of protein both released from damaged cells and produced as protective response (e.g., indicators of neuroinflammation). The identification of neurodamaging and/or neuroprotective components would help in the better understanding of the basic mechanisms involved in neurodegeneration and/or cell repair.

As expected, some proteins belonging to the acute phase response (α1-antitrypsin, haptoglobin 1-α1, -α2, and -β) showed an increased expression in severe TBI patients as a protective reaction that follow tissue injury-induced inflammation (Conti et al., 2004). In this study, careful consideration was given to the fact that CSF collection was done in TBI patients by an external ventricular catheter inserted as a part of normal therapeutic intervention to limit intracranial brain pressure, while in the control group CSF was collected by lumbar subarachnoid puncture. Indeed, it is known that a rostrocaudal concentration gradient of CSF proteins occurs in general, and that for some specific proteins, there is a concentration increase in the lumbar or ventricular CSF (Reiber, 2001). Thus, in order to minimize the effect of the rostrocaudal gradient, gels were normalized for the amount of protein loaded and differences evaluated on the basis of the relative abundance of proteins in CSF from TBI patients (ventricular) and control (lumbar). Moreover, since the regional (ventricular vs. lumbar) enrichment of specific proteins normally ranges from 1.5- to 3-fold, we considered relevant only protein spots showing more than 3-fold variation (increase or decrease) relative to control. In this study, more interesting was the exclusive finding

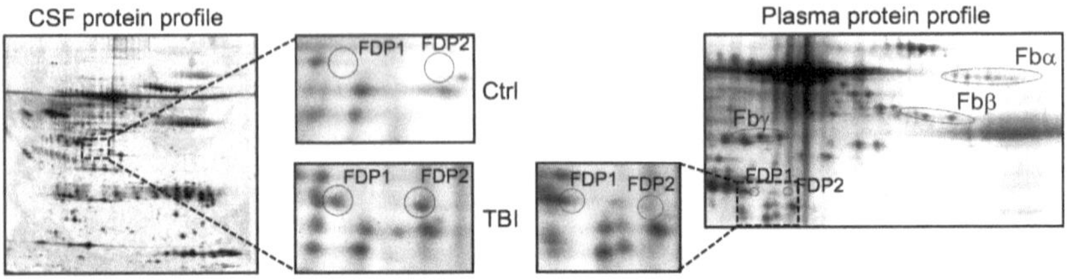

Figure 3 Comparison between 2DE map of CSF from severe traumatic brain injury (TBI) patients and control healthy subjects (Ctrl) reveals the specific presence of fibrinogen degradation products (FDP1 and FDP2) in the CSF of TBI patients. Analytical 2DE map of plasma from jugular vein bulb blood collected from TBI patients shows the presence of intact forms of different fibrinogens (α, β, γ) and the absence of FDPs, indicating that FDPs are generated in brain and were not the consequence of blood–brain barrier leakiness or blood sample contamination.

in TBI samples of two fibrinogen β degradation fragments (FDP1 and 2) (Fig. 3) that can be originated either by the fibrinogen degradation or by the degradation of its cross-linked fibrin products. Cross-linked fibrin degradation products (FDPs) have been detected in several neurological disorders (Brueton, Tugwell, Whittle, & Greenwood, 1974; Nowacki, Stankowska-Chomicz, & Borodin, 1990; Tovi & Nilsson, 1972). However, the mechanisms explaining the presence of FDPs in the CSF are not well understood; the specific contribution of increased permeability of the blood–CSF barrier, as opposed to the FDP local generation by fibrinolysis, is likely to be different in different pathological conditions. We found the presence of FDPs in the CSF of TBI patients in the absence of markers of blood–brain barrier breakage; in addition, the fragments were not present in the plasma of the same patients (Fig. 3), confirming the hypothesis of a local production by fibrinolytic activity. These types of fragments have also been found in CSF of patients with psychiatric and neurological disorders in the absence of traumatic lesion (Korschenhausen, Hampel, Ackenheil, Penning, & Muller, 1996; Okamoto, Takahashi, Minoura, Fujita, & Yoshimura, 1991; Wildenauer et al., 1991). In these disorders, the presence of the fragment has been thought due to local inflammation and in some cases correlated to the pathological events and to the clinical course (Okamoto et al., 1991; Wildenauer et al., 1991). The presence of the fragments in postmortem brain tissue and CSF of schizophrenic patients support that measurements in CSF from patients can reflect changes occurring in the CNS (Korschenhausen et al., 1996; Wildenauer et al., 1991).

Even if the role of these fragments or of the protein fibrinogen generation in response to TBI remains to determined, it is of interest that FDP1 expression in the CSF of severe TBI patients evaluated as spot optical density normalized for albumin content showed a significant correlations with Glasgow outcome scale measured at 6 months from trauma ($p=0.0167$, Spearman test $R=0.926$) (Table 1 and Fig. 4; unpublished observation). On the contrary, parameters defining the clinical conditions at the hospitalization soon after the trauma (i.e., Glasgow Coma Scale and Marshall Computed Tomography Classification) did not show correlation with FDPs (Table 1), suggesting that FDP1 might be considered as prognostic factor independent of the acute disease conditions.

Table 1 Clinical Features of Traumatic Brain Injury Patients and Presence of Fibrinogen Degradation Products in CSF

Patient	GCS	MCTC	GOS	FDP1[a]	FDP2[a]
2	7	3	3	139	150
1	7	3	4	256	322
3	7	1	4	282	184
6	6	5	4	439	689
4	7	5	5	543	507
5	3	5	5	648	526

[a]Spot optical density normalized to albumin.
GCS, Glasgow Coma Scale; MCTC, Marshall Computed Tomography Classification; GOS, Glasgow Outcome Scale; FDP, fragment degradation product.

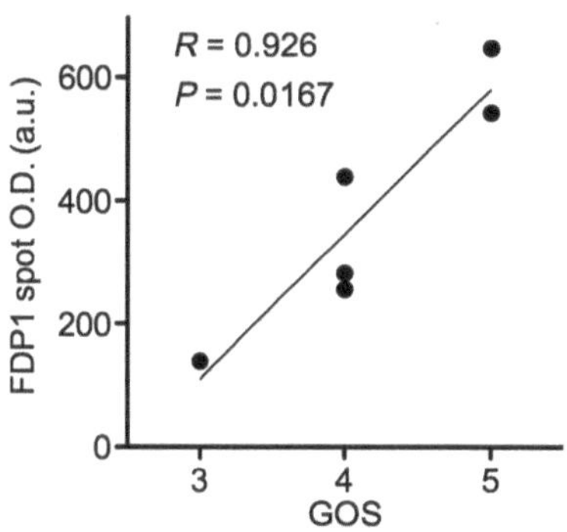

Figure 4 Fragment degradation product 1 (FDP1) expression in the CSF of severe traumatic brain injury patients correlates with Glasgow Outcome Scale (GOS) measured at 6 months from trauma. *R*, Spearman correlation value; O.D., optical density; a.u., arbitrary unit.

2.5.2 Characterization of PTMs in Pigment Epithelium-Derived Factor Isoforms Differentially Expressed in the CSF of Peripheral Neuropathies

In a comparative proteomics study aimed to identify relevant molecules characterizing the peripheral neuropathies either with or without pain, we found in the CSF three spots corresponding to the same protein, namely pigment epithelium-derived factor (PEDF), that is reduced in the pathological conditions compared to healthy subjects (Conti et al., 2005).

PEDF is a glycoprotein that belongs to the noninhibitory subclass of serpins (serine protease inhibitor) family and has been described as an important neurotrophic and neuroprotective factor (Tombran-Tink & Barnstable, 2003 and references therein). PEDF protects cerebellar granule neurons from induced apoptosis and embryonic hippocampal neurons from glutamate injury, as well as motor neurons from chronic glutamate-mediated degeneration (Bilak et al., 1999; DeCoster, Schabelman, Tombran-Tink, & Bazan, 1999; Nomura et al., 2001). It has also been reported that PEDF may work as niche signal for neuronal cell renewal (Ramirez-Castillejo et al., 2006), and its levels were found specifically increased in the CSF of individuals with ALS (Kuncl et al., 2002), suggesting that the elevated levels of PEDF may be an autoprotective response in ALS. Therefore, PEDF might have a role also in neurodegenerative diseases.

Interestingly, by Western blot performed with a specific antibody, it was possible to identify in the CSF the presence of several PEDF isoforms, at least six (Fig. 5), that may reflect different functional roles (Conti et al., 2005). The different electrophoretic migration properties of the six isoforms can be explained by the heterogeneity of the several PTMs described for the PEDF, namely the generation of pyroglutamate (pGlu) at the amino-terminus of the mature PEDF when the signal peptide is removed, the presence of one heterogeneous glycosidic residue that can include or not sialylation and fucosylation, and the possible phosphorylation on three serine residues (Fig. 5; Petersen, Valnickova, & Enghild, 2003). Thus, it was important to define the biochemical features of the three PEDF isoforms differentially expressed in peripheral neuropathy patients. This is an excellent example of how 2DE analysis can be fundamental in dissecting PTMs of specific proteins.

The generation of pGlu blocking group at the amino-terminus is a PTM important for PEDF function, because it prevents protein degradation by most of aminopeptidases, increasing protein half-life in biological fluids. This regulates the function of PEDF that can be activated by pGlu removal,

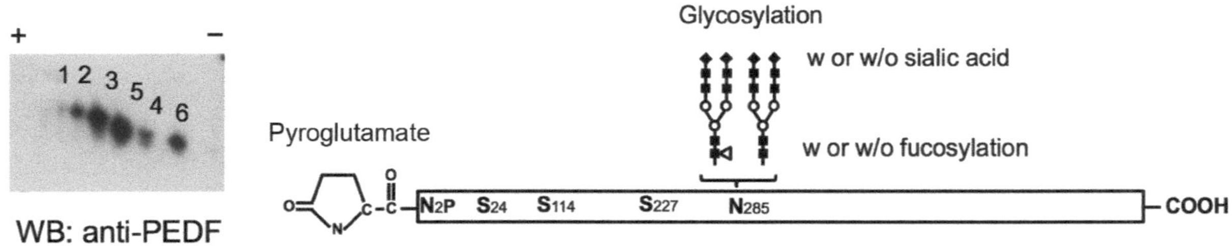

Figure 5 2DE pattern of pigment epithelium-derived factor (PEDF) isoforms and schematic representation of the possible PEDF post-translation modifications.

where and when necessary, by secreted proteases belonging to the family of pyroglutamyl peptidases (Dando, Fortunato, Strand, Smith, & Barrett, 2003). From the electrophoretic mobility point of view, the generation of pGlu prevents the protonation of amino-terminus of the protein resulting in a more acidic net charge; thus, the observation that removal of pGlu, by treatment with pGlu-aminopeptidase, resulted in a shift of all six PEDF isoforms to more basic pH (as the result of amino-terminus group protonation) indicated that pGlu was not the peculiar feature of the differentially expressed PEDF isoforms (Fig. 6, top-right panels; Conti et al., 2005). Then,

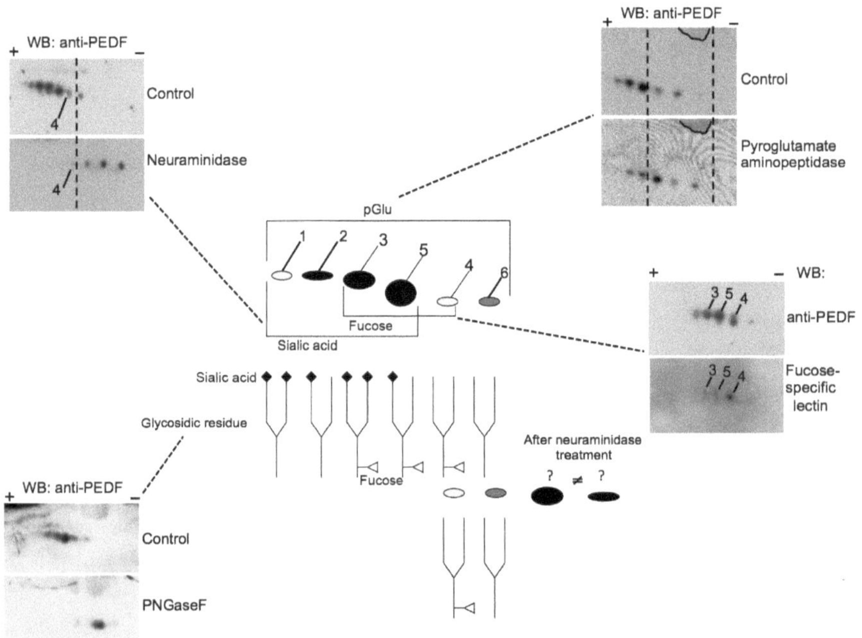

Figure 6 Schematic representation of the pigment epithelium-derived factor (PEDF) isoforms and their posttranslational modifications. Isoforms 3, 5, and 4 are differentially expressed in the CSF of peripheral neuropathies. All six isoforms contain pyroglutamate (pGlu), as inferred by treatment with pGlu-aminopeptidase that produces a spot shift to basic area of the 2DE (top-right panels), and are glycosylated, as inferred by reduction in molecular weight and shift to basic area of the gel induced by PNG-aseF treatment (bottom-left panels). The more acidic isoforms (1, 2, 3, and 5) contain sialic acid, while isoforms 4 and 6 are apparently nonsialylated, as inferred by sialic acid removal by treatment with neuraminidase (top-left panels). The differentially expressed isoforms are characterized by the presence of fucose residue on the complex oligosaccharide, as indicated by the reactivity of the 3, 5, and 4 PEDF isoforms with the fucose-specific lectin from *Tetragonolobus purpureas*.

we investigated whether protein glycosylation may be important for different electrophoretic migration features. CSF treatment with PNGase F demonstrated that PEDF contained complex oligosaccharides that may be taken into account for heterogeneity (Fig. 6, bottom-left panels; Conti et al., 2005). Removal from glycosidic residue of sialic acid groups by neuraminidase treatment resulted in a reduction of PEDF isoforms' heterogeneity (from 6 to 4 spots) and in a shift to more basic pH of some isoforms (Fig. 6, top-left panels). Apparently, two isoforms seemed not to be affected by the treatment, suggesting the absence of sialic acid in these protein spots that did not correspond to the differentially expressed isoforms (Fig. 6). Thus, also protein sialylation was not the unique feature of the differentially expressed PEDF isoforms, even though different rate of sialylation might affect the interaction with the putative receptor or with collagens and glycosaminoglycans, supporting different PEDF functional roles. Then, we exploited the reactivity of *Tetragonolobus purpureas* agglutinin, a lectin that specifically binds to α-L-fucose, to define the oligosaccharide composition of the different PEDF isoforms. Three of six PEDF isoforms reacted with the fucose-specific lectin, indicating that the differentially expressed isoforms are the fucosylated one (Fig. 6, bottom-right panels). This is an important feature because fucose residues play a fundamental role in protein–protein and protein–extracellular matrix interaction and might regulate the already described interactions of PEDF with extracellular matrix component such as collagens and glycosaminoglycans (Alberdi, Weldon, & Becerra, 2003; Yasui et al., 2003). The residual PEDF heterogeneity observed after sialic acids removal might be ascribed to serine residues phosphorylation that had not been taken into consideration at the time of the study because they had not yet described (Maik-Rachline, Shaltiel, & Seger, 2005).

2.5.3 Skeletal Muscle Involvement in ALS: Analysis of Protein Pattern Downstream to the Differentially Expressed Myosin-Binding Protein H

In ALS, skeletal muscle is considered the final target of motor neuron degeneration; nevertheless, recent findings suggest that skeletal muscle might be actively involved in the pathogenesis of the disease (Musaro, 2010): in an ALS animal model, for example, a progressive muscle atrophy is one of the earliest detectable events prior to neuromuscular junction alterations, retrograde axonal degeneration, and motor neuron degeneration (Dobrowolny, Aucello, & Musaro, 2011; Dobrowolny et al., 2008; Wong & Martin, 2010).

A differential proteomic approach has been applied to investigate modifications occurring in proteins of the skeletal muscle at the onset of the disease in sporadic ALS patients, compared to that of healthy controls, motor neuropathies, and myopathy patients (Conti et al., 2014). The main finding of this work was that among several muscle-specific proteins that were downregulated (e.g., myosin light chain, tropomyosin, actin, etc.), the skeletal muscles of ALS patients showed increased expression of the myosin-binding protein H (MyBP-H), a feature able to distinguish ALS patients from motor neuropathies already at the onset of the disease when differential diagnosis is necessary since clinical symptoms are not yet defined (Conti et al., 2014). MyBP-H is a component of the thick filaments of the skeletal muscle with strong affinity for myosin; however, its function is still unknown (Gilbert, Cohen, Pardo, Basu, & Fischman, 1999). In a cell line, MyBP-H has been reported to interact and regulate the Rho kinase 1 (ROCK1), which in turn affects actin–myosin interaction and cell motility (Hosono et al., 2011). The ROCK pathway regulates actin–myosin filament assembly by phosphorylation of numerous substrate proteins; among these, the LIM kinases (LIMK), whose specific activity on cofilin inactivates actin depolymerization function, in turn result in actin filament stabilization (Olson, 2008). Given the increased expression of MyBP-H in ALS muscle, we investigated the same samples whether the downstream pattern of molecules along the ROCK pathway was also affected. Even though we detected an increased expression of ROCK2, the ROCK isoform expressed in muscle, the downstream signaling was found to be impaired. In fact, treatment with lambda phosphatase (λ-PPase) disclosed the accumulation of a nonphosphorylated isoform of LIMK1 in ALS patients compared to control healthy subjects, and the less acidic 2DE pattern of cofilin2 indirectly suggested that, downstream to LIMK1, cofilin2 was also less phosphorylated in ALS patients (Fig. 7; Conti et al., 2014). Therefore, following the differential proteomics analysis, the further hypothesis-driven investigations, that were carried out, clearly suggest that in ALS, in addition to anomalous MyBP-H expression, the downstream protein pattern of molecules involved in muscular actin–myosin regulation is also affected in pathological conditions.

The relevance of the increased expression of MyBP-H observed in ALS patients is supported both at mRNA and at protein levels in motor neuron degeneration animal models. For example, increased MyBP-H mRNA has been reported in the muscle of $SOD1^{G86R}$ transgenic mice ALS model in the early stage of disease onset, and in mice 7 days after sciatic nerve axotomy, prior to motor neuron death (Gonzalez de Aguilar et al., 2008).

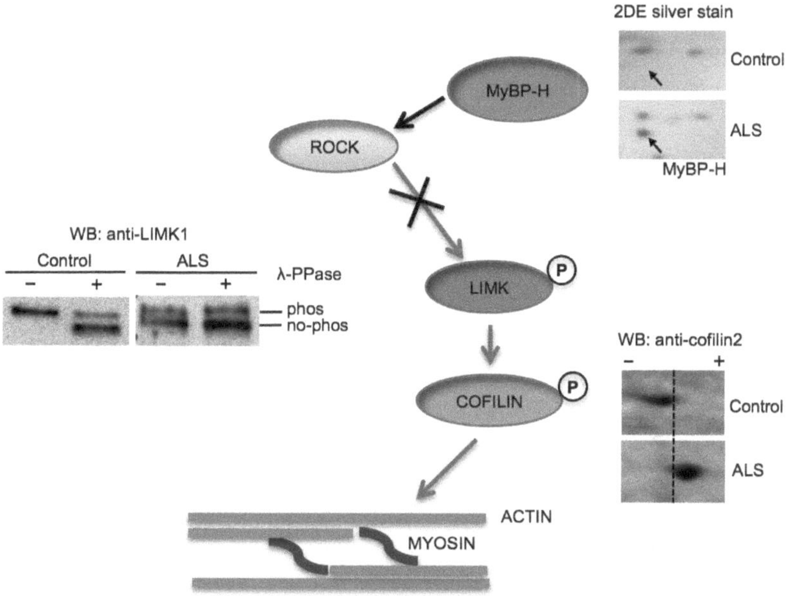

Figure 7 Schematic representation shows how MyBP-H might affect actomyosin organization through interaction with ROCK, which in turn leads to inhibition of the LIMK–cofilin phosphorylation pathway. 2DE silver staining shows the increased expression of MyBP-H in ALS muscle. Western blot (WB) of anti-LIMK2 shows that a nonphosphorylated lower band, which induced by treatment with λ-phosphatase (λPPase) in control subjects, is already present in nontreated ALS patients. WB of anti-cofilin2 shows a less acidic, likely nonphosphorylated, spot in ALS patients.

At the protein level, a transitory increase in MyBP-H expression has been reported in the muscle of 7-week-old $SOD1^{G93A}$ transgenic mice model for ALS, when overt clinical symptoms at hindlimbs have not yet occurred, and then MyBP-H expression decreased at 14-week-old mice exclusively in the muscles of pathologically compromised hindlimbs, but not in the muscles of the not yet compromised forelimbs (Capitanio et al., 2012). Furthermore, in the muscles of a rat model of pharmacologically induced postsynaptic block of neuromuscular transmission, an increased mRNA expression of MyBP-H was reported concomitantly with a downregulation of all other myofibrillar proteins (Norman et al., 2006). On the contrary, when muscular atrophy occurs in the absence of neuromuscular junction alteration (e.g., in response to food deprivation), MyBP-H expression was reported to decrease (Jagoe, Lecker, Gomes, & Goldberg, 2002). These findings suggest that MyBP-H upregulation might represent an early sign of muscle regeneration that is consistent with a model whereby muscular alteration is an early pathological event in ALS, whether this event is associated with neuromuscular junction alteration and is independent from motor neuron degeneration remains to be defined.

2.5.4 Expression of Nonsialylated Ceruloplasmin in the CSF of ALS Patients

ALS like some other neurodegenerative diseases (e.g., PD and AD) is characterized by oxidative stress, the condition arising from an imbalance between the physiological production of potentially toxic ROS (i.e., superoxide, hydrogen peroxide, and hydroxyl radical), and the scavenging activities. These protective actions include enzymatic activities (as superoxide dismutase), low molecular weight antioxidant species (as vitamin E), and other forms of protection such as systems for metal transport (Carri, Ferri, Cozzolino, Calabrese, & Rotilio, 2003). In the case of metal transport systems, the correct handling of transition metal ions, like copper and iron, is crucial since these metals are able to undergo redox cycles, during which toxic ROS are generated.

One of these metal transporters is the copper–protein ceruloplasmin (Cp), secreted by choroid plexus in the CSF and expressed as membrane-bound glycosylphosphatidylinositol (GPI)-anchored isoform by astrocytes and by leptomeningeal cells in the CNS (Mittal, Doroudchi, Jeong, Patel, & David, 2003). Cp acts mainly as ferroxidase, oxidizing toxic ferrous iron to nontoxic ferric form (Hellman & Gitlin, 2002), playing an important role in brain's iron homeostasis and antioxidant defense (Texel, Xu, & Harris, 2008). The importance of the protective function of Cp is demonstrated by aceruloplasminemia patients, showing massive brain iron deposition, and for this reason, aceruloplasminemia has been included in the group of diseases so-called NBIA (neurodegeneration with brain iron accumulation) (Harris, Klomp, & Gitlin, 1998; Kono, 2012). Furthermore, Cp-null mice show iron accumulation in the CNS and an increased free radical-mediated damage (Patel et al., 2002). It is therefore possible that changes in CSF Cp expression level and/or in protein modification(s), which can affect enzymatic activity, may contribute to the neurodegenerative process exacerbating the oxidative damage. Thus, we performed an hypothesis-driven comparative proteomics analysis of the Cp's features in the CSF of some neurodegenerative diseases.

In the CSF from ALS patients, we found an increased expression of Cp compared to both healthy subjects and peripheral neuropathy patients; interestingly, 2DE revealed that the expression increase was mainly referred to the more basic isoform of Cp (Fig. 8; Conti et al., 2008). Comparative

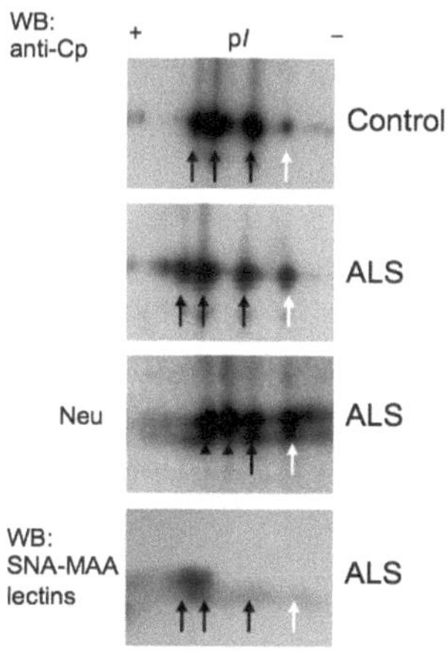

Figure 8 Comparison between 2DE pattern profiles of the ceruloplasmin (Cp) expressed in the CSF from healthy subjects (control) and patients affected by amyotrophic lateral sclerosis (ALS) reveals increased expression of the more basic Cp isoform (ALS, white arrows). This isoform corresponds to the nonsialylated Cp as indicated by signal accumulation upon treatment with neuraminidase (Neu) and by the lack of reactivity with sialic acid-specific lectins *Sambucus nigra* (SNA) and *Maackia amurensis* (MAA).

analysis of Cp 2DE pattern with either the serum form of Cp or the GPI-linked isoform excluded that the increase of the basic isoform of Cp in the CSF of ALS patients was the result of a leakiness of the blood–CSF barrier or the results of regulated release of the GPI-membrane-bound form of Cp, respectively (Conti et al., 2008). Then the analysis of the glycosidic heterogeneity of Cp showed that the more basic isoform of Cp corresponded to the nonsialylated protein, as inferred by the shift to the more basic isoform of Cp upon neuraminidase treatment that remove sialic acid, and by the lack of reactivity of the more basic Cp isoform with the sialic acid-specific lectins (*Sambucus nigra*, SNA and *Maackia amurensis*, MAA) (Fig. 8; Conti et al., 2008). Despite the increased expression in basic isoform of Cp, the total CSF ferroxidase activity in ALS patients was comparable to that of the controls, suggesting a Cp functional impairment (Conti et al., 2008). The increase in asialo-Cp may be the result of an isoform interconversion due to endogenous sialidase activity. Interestingly, a general reduction in the content of sialic acid in CSF glycoprotein has been reported in ALS patient (Niebroj-Dobosz, Janik, Jamrozik, & Kwiecinski, 1999). Whether the sialic acid release is a regulated phenomenon or is a side effect is not known; however, its removal from Cp may have a detrimental effect on ALS, because changes in carbohydrate structure have functional implication on protein turnover and enzymatic activity. In fact, while it is not clear if ferroxidase activity is affected in asialo-Cp, it is clear that the asialo-Cp is quickly removed from circulation by liver uptake (Morell, Gregoriadis, Scheinberg, Hickman, & Ashwell, 1971; Morell, Irvine, Sternlieb, Scheinberg, & Ashwell, 1968). Therefore, the Cp's increased expression could be the balancing response of the organism to overcome the partial inactivation of ferroxidase function and to counteract the increased removal of asialo-Cp.

2.5.5 Oxidation of Ceruloplasmin in the CSF from PD and AD Promotes Loss of Ferroxidase Function

The differential analysis of 2DE expression pattern of Cp in neurodegenerative disorders characterized by oxidative stress and CNS iron deposition has also been performed in the CSF from PD and AD (Ward, Zucca, Duyn, Crichton, & Zecca, 2014). On the contrary of what was observed for ALS patients, in AD and PD, Cp did not change its expression level in CSF, but showed the presence of more acidic isoforms (Fig. 9) that are able to statistically discriminate PD and AD patients both from healthy subjects and from patients with different neurological diseases (Olivieri et al., 2011).

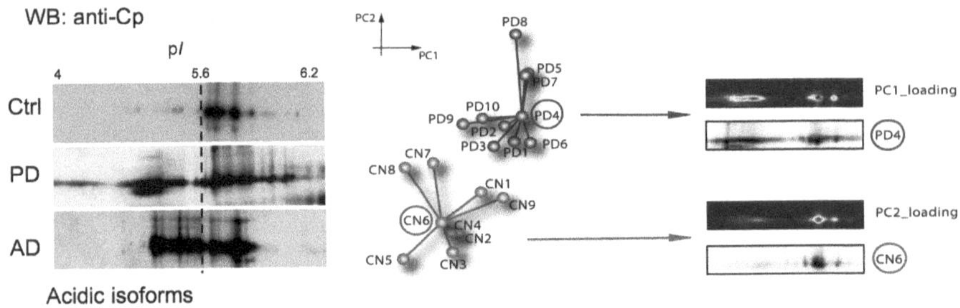

Figure 9 The differential analysis of 2DE expression pattern of ceruloplasmin (Cp) in the CSF of neurodegenerative disorders characterized by oxidative stress showed the presence of more acidic isoforms of Cp in Parkinson's disease (PD) and Alzheimer's disease (AD) patients compared to control healthy subjects (Ctrl). Unsupervised dimensional reduction analysis, performed by principal component analysis, followed by cluster identification approach applied to the Cp electrophoretic patterns identify two distinct groups corresponding to PD patients (blue (gray in the print version) dots) and control healthy subjects (red (light gray in the print version) dots). False color images represent the signals characterizing the PD cluster (PC1) and healthy subjects' cluster (PC2), respectively, while PD4 and CN6 are the real sample images representing the centroid-prototype of the cluster.

By using an unsupervised biocomputational automatic cluster identification approach (Cannistraci et al., 2010) applied to the Cp Western blot images, it was possible to discriminate both PD patients from control subjects (Fig. 9) and PD from AD patients, on the basis of the Cp pattern profile (Olivieri et al., 2011).

The acidification of Cp's isoelectric point is due to oxidative modifications as inferred by the observation that *in vitro* oxidation of CSF generated a Cp shift resembling that observed in PD and AD patients, and by the fact that acidic Cp isoforms showed to be carbonylated, a modification that is induced by oxidative conditions and that can be detected by derivatization with 2,4-dinitrophenilhydrazine and 2DE Western blot (Fig. 10; Olivieri et al., 2011). The prooxidant condition that occurs in neurodegenerative disorder CSF is underlined by the observed general increase of total protein carbonylation in CSF from PD (Iannaccone et al., 2013) and by the observation that exogenous purified Cp added and *ex vivo* aged for 9 days in CSF from AD patients resulted in oxidative modifications that did not occurred if aging was done in CSF of healthy subjects (Barbariga et al., 2014). Therefore, PD and AD pathological CSF contain compounds/small molecules that foster protein oxidative modifications, likely contributing to exacerbate the disorders. In the case of the analysis of post-translational oxidative modifications, special attention must be given to the risk of artifactual modification due to the *ex vivo* protein oxidation fostered by the experimental conditions or by the sample storage conditions (Olivieri et al., 2011; Poulsen et al., 2014). In our experience, we avoid experimental oxidation by storing the sample and performing the

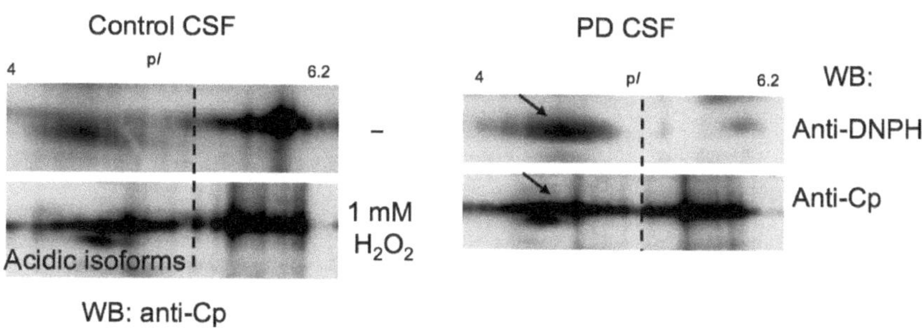

Figure 10 *In vitro* oxidative stress, induced in the CSF from control healthy subject by treatment with 1 mM of H_2O_2, fosters Cp to convert to acidic isoforms (panels on the left). Acidic Cp isoforms present in the CSF from PD patients are modified by carbonylation as revealed by derivatization with 2,4-dinitrophenilhydrazine (DNPH) followed by Western blot with anti-DNPH antibody (panels on the right).

experiments under nitrogen-conditioned atmosphere (Barbariga et al., 2014; Olivieri et al., 2011).

Interestingly, the observed oxidation rate of Cp correlated with UPDRS clinical grading in PD patients, and from the mechanistic point of view, the Cp oxidation resulted in the enzymatic loss of ferroxidase activity which in turn promoted intracellular iron accumulation in neurons and astrocytes (Olivieri et al., 2011).

Thus, the changes in electrophoretic pattern observed in CSF Cp from PD and AD patients might reflect accelerated protein aging, as induced by oxidative pathological conditions (Grimm, Hoehn, Davies, & Grune, 2011), and might be considered as a potential marker for the evaluation of oxidative stress levels in the CNS of patients. Furthermore, the results indicated that Cp oxidation might play a role in pathological mechanisms that promote intracellular iron retention in CNS.

2.5.6 Oxidation-Induced Structural Changes Foster Ceruloplasmin Deamidation with Gain of Integrin-Binding Function

Another PTM that can be promoted by the oxidative environment of neurodegenerative diseases is protein deamidation.

Asparagine deamidation is a spontaneous chemical reaction that leads to the formation of aspartate and isoaspartate (isoAsp) in proteins, and occurs *in vivo* during protein aging (Weintraub & Deverman, 2007). The rate of Asn deamidation is determined by several factors, including protein sequence and secondary/tertiary structure, the pH, ionic strength, and temperature (Weintraub & Deverman, 2007). Asn deamidation causes a change in the peptide backbone length that can alter protein function and stability and render the proteins more susceptible to degradation. Thus, Asn deamidation is considered as a sort of molecular timer that marks the spontaneous protein aging (Weintraub & Deverman, 2007). In pathological conditions, Asn deamidation has been associated with oxidative stress (Shimizu, Matsuoka, & Shirasawa, 2005; Watanabe, Takio, & Ihara, 1999), like in AD where the deamidated forms of Tau protein and β-amyloid peptides accumulate in pathological aggregates (Hasegawa et al., 1992; Shimizu, Watanabe, Ogawara, Mori, & Shirasawa, 2000). Asn deamidation may also have biological function consequences; indeed, it has been shown that the tripeptide Asparagine–Glycine–Arginine (NGR) motif can undergo deamidation that results in the formation of isoDGR (isoAsp-Gly-Arg), a motif that is able to recognize the RGD-binding site of integrins (Corti & Curnis, 2011). From the methodological point of view, in the case of protein deamidation analysis, it is mandatory to modify the standard protocol used for tryptic digestion before MS because the usual conditions (ammonium bicarbonate buffer at pH 8.5, at 37 °C) strongly induce *in vitro* peptides deamidation (Barbariga et al., 2014). Thus, many time, depending on the type of modification analyzed, the methods should be modified or *de novo* developed as to generate simple, robust, and reproducible protocols aimed to avoid the generation of artificial modifications.

The Cp sequence contains two NGRs motifs, one (^{568}NGR) is exposed on the surface of the protein and the other (^{962}NGR) hidden inside its 3D structure, and thus less accessible to solvents (Barbariga et al., 2014; Fig. 11). Since oxidative conditions might cause structural changes that accelerate Asn deamidation, we investigated whether the Cp NGR motifs can deamidate during aging under oxidative conditions, the protein aging condition that we hypothesized to occur in pathological CSF. We found that both Cp NGR sites can deamidate; however, while ^{568}NGR is free to convert to iso-DGR (Fig. 11), the ^{962}NGR motif undergoes deamidation only when Asn aging occurs under oxidative conditions, in which Cp structural changes are induced, as demonstrated by circular dichroism analysis (Barbariga et al., 2014). The NGR to isoDGR transition in Cp results in a loss of ferroxidase activity and a gain in integrin-binding function (Barbariga et al., 2014). Protein oxidation and conformational changes can cause release from Cp of the copper ions that are necessary for the Cp ferroxidase activity (Musci, Bonaccorsi di Patti, Fagiolo, & Calabrese, 1993; Olivieri et al., 2011; Sedlak, Zoldak, & Wittung-Stafshede, 2008). The deamidation of NGR

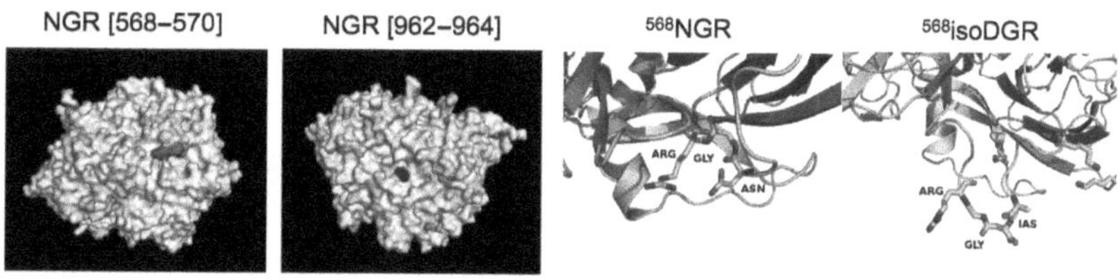

Figure 11 Structural analysis of Cp's NGR motifs. Panels on the left: surface representation of Cp, in red (light gray in the print version) and blue (gray in the print version) are highlighted the ^{568}NGR and ^{962}NGR sequences, respectively, as calculated by the GROMACS g_sas tool on crystallographic Cp structure (Protein Data Bank code 2J5W). The higher solvent-accessible surface of ^{568}NGR is represented by the larger size of the colored region, while ^{962}NGR is buried inside Cp structure. Panels on the right: structural comparison of the molecular dynamics modeling of the surface expose ^{568}NGR and its deamidated counterpart ^{568}isoDGR.

motifs to isoDGR induces a gain of a new Cp biological function in terms of binding to integrins and in the ability to trigger an intracellular signaling cascade, as indicated by reverse phase protein array analysis, that collectively may regulate cell cycle, proliferation, and cytoskeletal rearrangement (Barbariga et al., 2014). Therefore, Cp deamidation can produce a biologically relevant gain of function, as also suggested by the fact that the NGR sites of Cp are very high conserved sequences across differing species from hamster to human. Interestingly, we demonstrated that *ex vivo* aging of purified Cp in the CSF from AD patients impairs Cp ferroxidase activity and favors NGR motifs' deamidation, which in turn promotes integrin-binding properties (Barbariga et al., 2014). This implied the presence in pathological CSF of factors that foster/accelerate deamidation either directly or as consequence of the induction of structural changes, and that these modifications might occur *in vivo* in pathological conditions. Several cell types in the CNS that express integrins on their surface can be in principle the target of deamidated-Cp, including neurons, astrocytes, and microglia cells. Anyway, one of the interesting potential targets of deamidated-Cp is the specialized epithelial cells of the ependymal layer and choroid plexus, which are directly in contact with the CSF and have been reported to be altered, for example, in AD (Krzyzanowska & Carro, 2012; Perez-Gracia, Blanco, Carmona, Carro, & Ferrer, 2009; Serot, Zmudka, & Jouanny, 2012). In conclusion, the Cp alterations observed in PD and AD patients (Olivieri et al., 2011) might reflect accelerated protein aging, as a consequence of changes in the environmental redox status of pathological CSF that foster both Cp oxidation and Asn deamidation of NGR motifs, which in turn result in the loss of ferroxidase function and in a gain in integrin-binding properties.

3. CONCLUSIONS

In the light of our experience, even if it is clear that most of the proteomics studies performed in the field of neurodegenerative diseases have been done to disclose candidate biomarkers, at the present time it is difficult to conclude that they have been successful, and there is still the need of standardized validation studies. More useful are results obtained by proteomics-based approaches aimed to characterize specific PTMs that in some cases resulted to be very informative in disclosing mechanisms that can contribute to the pathology and that might be the target of innovative therapeutic strategies.

REFERENCES

Alberdi, E. M., Weldon, J. E., & Becerra, S. P. (2003). Glycosaminoglycans in human retinoblastoma cells: Heparan sulfate, a modulator of the pigment epithelium-derived factor-receptor interactions. *BMC Biochemistry*, *4*, 1.

Alberio, T., Bossi, A. M., Milli, A., Parma, E., Gariboldi, M. B., Tosi, G., et al. (2010). Proteomic analysis of dopamine and alpha-synuclein interplay in a cellular model of Parkinson's disease pathogenesis. *The FEBS Journal*, *277*(23), 4909–4919.

Almeras, L., Lefranc, D., Drobecq, H., de Seze, J., Dubucquoi, S., Vermersch, P., et al. (2004). New antigenic candidates in multiple sclerosis: Identification by serological proteome analysis. *Proteomics*, *4*(7), 2184–2194.

Andreev, V. P., Petyuk, V. A., Brewer, H. M., Karpievitch, Y. V., Xie, F., Clarke, J., et al. (2012). Label-free quantitative LC-MS proteomics of Alzheimer's disease and normally aged human brains. *Journal of Proteome Research*, *11*(6), 3053–3067.

Barbariga, M., Curnis, F., Spitaleri, A., Andolfo, A., Zucchelli, C., Lazzaro, M., et al. (2014). Oxidation-induced structural changes of ceruloplasmin foster NGR-motifs deamidation that promote integrin binding and signalling. *The Journal of Biological Chemistry*, *289*(6), 3736–3748.

Bilak, M. M., Corse, A. M., Bilak, S. R., Lehar, M., Tombran-Tink, J., & Kuncl, R. W. (1999). Pigment epithelium-derived factor (PEDF) protects motor neurons from chronic glutamate-mediated neurodegeneration. *Journal of Neuropathology and Experimental Neurology*, *58*(7), 719–728.

Blennow, K. (2005). CSF biomarkers for Alzheimer's disease: Use in early diagnosis and evaluation of drug treatment. *Expert Review of Molecular Diagnostics*, *5*(5), 661–672.

Bossuyt, P. M., Reitsma, J. B., Bruns, D. E., Gatsonis, C. A., Glasziou, P. P., Irwig, L. M., et al. (2003). The STARD statement for reporting studies of diagnostic accuracy: Explanation and elaboration. *Clinical Chemistry*, *49*(1), 7–18.

Bradley, B. P. (2012). Finding biomarkers is getting easier. *Ecotoxicology*, *21*(3), 631–636.

Brueton, M. J., Tugwell, P., Whittle, H. C., & Greenwood, B. M. (1974). Fibrin degradation products in the serum and cerebrospinal fluid of patients with group A meningococcal meningitis. *Journal of Clinical Pathology*, *27*(5), 402–404.

Buono, P., D'Armiento, F. P., Terzi, G., Alfieri, A., & Salvatore, F. (2001). Differential distribution of aldolase A and C in the human central nervous system. *Journal of Neurocytology*, *30*(12), 957–965.

Butterfield, D. A., Reed, T. T., Perluigi, M., De Marco, C., Coccia, R., Keller, J. N., et al. (2007). Elevated levels of 3-nitrotyrosine in brain from subjects with amnestic mild cognitive impairment: Implications for the role of nitration in the progression of Alzheimer's disease. *Brain Research*, *1148*, 243–248.

Butterfield, D. A., & Sultana, R. (2008a). Identification of 3-nitrotyrosine-modified brain proteins by redox proteomics. *Methods in Enzymology*, *440*, 295–308.

Butterfield, D. A., & Sultana, R. (2008b). Redox proteomics: Understanding oxidative stress in the progression of age-related neurodegenerative disorders. *Expert Review of Proteomics*, *5*(2), 157–160.

Cannistraci, C. V., Alanis-Lobato, G., & Ravasi, T. (2013a). From link-prediction in brain connectomes and protein interactomes to the local-community-paradigm in complex networks. *Scientific Reports*, *3*, 1613.

Cannistraci, C. V., Alanis-Lobato, G., & Ravasi, T. (2013b). Minimum curvilinearity to enhance topological prediction of protein interactions by network embedding. *Bioinformatics*, *29*(13), i199–i209.

Cannistraci, C. V., Montevecchi, F. M., & Alessio, M. (2009). Median-modified Wiener filter provides efficient denoising, preserving spot edge and morphology in 2-DE image processing. *Proteomics*, *9*(21), 4908–4919.

Cannistraci, C. V., Ravasi, T., Montevecchi, F. M., Ideker, T., & Alessio, M. (2010). Nonlinear dimension reduction and clustering by minimum curvilinearity unfold neuropathic pain and tissue embryological classes. *Bioinformatics*, *26*, i531–i539.

Capitanio, D., Vasso, M., Ratti, A., Grignaschi, G., Volta, M., Moriggi, M., et al. (2012). Molecular signatures of amyotrophic lateral sclerosis disease progression in hind and forelimb muscles of an SOD1(G93A) mouse model. *Antioxidants & Redox Signaling*, *17*(10), 1333–1350.

Caron, M., Choquet-Kastylevsky, G., & Joubert-Caron, R. (2007). Cancer immunomics using autoantibody signatures for biomarker discovery. *Molecular & Cellular Proteomics: MCP*, *6*(7), 1115–1122.

Carri, M. T., Ferri, A., Cozzolino, M., Calabrese, L., & Rotilio, G. (2003). Neurodegeneration in amyotrophic lateral sclerosis: The role of oxidative stress and altered homeostasis of metals. *Brain Research Bulletin*, *61*(4), 365–374.

Castano, E. M., Roher, A. E., Esh, C. L., Kokjohn, T. A., & Beach, T. (2006). Comparative proteomics of cerebrospinal fluid in neuropathologically-confirmed Alzheimer's disease and non-demented elderly subjects. *Neurological Research*, *28*(2), 155–163.

Castegna, A., Aksenov, M., Aksenova, M., Thongboonkerd, V., Klein, J. B., Pierce, W. M., et al. (2002). Proteomic identification of oxidatively modified proteins in Alzheimer's disease brain. Part I: Creatine kinase BB, glutamine synthase, and ubiquitin carboxy-terminal hydrolase L-1. *Free Radical Biology & Medicine*, *33*(4), 562–571.

Choi, J., Levey, A. I., Weintraub, S. T., Rees, H. D., Gearing, M., Chin, L. S., et al. (2004). Oxidative modifications and down-regulation of ubiquitin carboxyl-terminal hydrolase L1 associated with idiopathic Parkinson's and Alzheimer's diseases. *The Journal of Biological Chemistry*, *279*(13), 13256–13264.

Church, A. J., Cardoso, F., Dale, R. C., Lees, A. J., Thompson, E. J., & Giovannoni, G. (2002). Anti-basal ganglia antibodies in acute and persistent Sydenham's chorea. *Neurology*, *59*(2), 227–231.

Conti, A., Iannaccone, S., Sferrazza, B., De Monte, L., Cappa, S., Franciotta, D., et al. (2008). Differential expression of ceruloplasmin isoforms in the cerebrospinal fluid of amyotrophic lateral sclerosis patients. *Proteomics. Clinical Applications*, *2*(12), 1628–1637.

Conti, A., Ricchiuto, P., Iannaccone, S., Sferrazza, B., Cattaneo, A., Bachi, A., et al. (2005). Pigment epithelium-derived factor is differentially expressed in peripheral neuropathies. *Proteomics*, *5*(17), 4558–4567.

Conti, A., Riva, N., Pesca, M., Iannaccone, S., Cannistraci, C. V., Corbo, M., et al. (2014). Increased expression of myosin binding protein H in the skeletal muscle of amyotrophic lateral sclerosis patients. *Biochimica et Biophysica Acta*, *1842*(1), 99–106.

Conti, A., Sanchez-Ruiz, Y., Bachi, A., Beretta, L., Grandi, E., Beltramo, M., et al. (2004). Proteome study of human cerebrospinal fluid following traumatic brain injury indicates fibrin(ogen) degradation products as trauma-associated markers. *Journal of Neurotrauma*, *21*(7), 854–863.

Corti, A., & Curnis, F. (2011). Isoaspartate-dependent molecular switches for integrin-ligand recognition. *Journal of Cell Science*, *124*(Pt. 4), 515–522.

Craft, G. E., Chen, A., & Nairn, A. C. (2013). Recent advances in quantitative neuroproteomics. *Methods*, *61*(3), 186–218.

Cumming, R. C., & Schubert, D. (2005). Amyloid-beta induces disulfide bonding and aggregation of GAPDH in Alzheimer's disease. *FASEB Journal: Official Publication of the Federation of American Societies for Experimental Biology*, *19*(14), 2060–2062.

Dale, R. C., Heyman, I., Giovannoni, G., & Church, A. W. (2005). Incidence of anti-brain antibodies in children with obsessive-compulsive disorder. *The British Journal of Psychiatry*, *187*, 314–319.

Dando, P. M., Fortunato, M., Strand, G. B., Smith, T. S., & Barrett, A. J. (2003). Pyroglutamyl-peptidase I: Cloning, sequencing, and characterisation of the recombinant human enzyme. *Protein Expression and Purification, 28*(1), 111–119.

D'Ascenzo, M., Relkin, N. R., & Lee, K. H. (2005). Alzheimer's disease cerebrospinal fluid biomarker discovery: A proteomics approach. *Current Opinion in Molecular Therapeutics,* 7(6), 557–564.

Davidsson, P., & Sjogren, M. (2005). The use of proteomics in biomarker discovery in neurodegenerative diseases. *Disease Markers, 21*(2), 81–92.

Davidsson, P., Westman-Brinkmalm, A., Nilsson, C. L., Lindbjer, M., Paulson, L., Andreasen, N., et al. (2002). Proteome analysis of cerebrospinal fluid proteins in Alzheimer patients. *Neuroreport, 13*(5), 611–615.

De Monte, L., Sanvito, F., Olivieri, S., Vigano, F., Doglioni, C., Frasson, M., et al. (2008). Serological immunoreactivity against colon cancer proteome varies upon disease progression. *Journal of Proteome Research,* 7(2), 504–514.

DeCoster, M. A., Schabelman, E., Tombran-Tink, J., & Bazan, N. G. (1999). Neuroprotection by pigment epithelial-derived factor against glutamate toxicity in developing primary hippocampal neurons. *Journal of Neuroscience Research, 56*(6), 604–610.

Dobrowolny, G., Aucello, M., & Musaro, A. (2011). Muscle atrophy induced by SOD1G93A expression does not involve the activation of caspase in the absence of denervation. *Skeletal Muscle, 1*(1), 3.

Dobrowolny, G., Aucello, M., Rizzuto, E., Beccafico, S., Mammucari, C., Boncompagni, S., et al. (2008). Skeletal muscle is a primary target of SOD1G93A-mediated toxicity. *Cell Metabolism, 8*(5), 425–436.

El-Fawal, H. A. N. (2014). Neuroantibody biomarkers: Links and challenges in environmental neurodegeneration and autoimmunity. *Autoimmune Diseases, 2014*, 340875. http://dx.doi.org/10.1155/2014/340875.

Fasano, M., Alberio, T., Colapinto, M., Mila, S., & Lopiano, L. (2008). Proteomics as a tool to investigate cell models for dopamine toxicity. *Parkinsonism & Related Disorders, 14,* S135–S138.

Ferrer, I., Martinez, A., Blanco, R., Dalfo, E., & Carmona, M. (2011). Neuropathology of sporadic Parkinson disease before the appearance of parkinsonism: Preclinical Parkinson disease. *Journal of Neural Transmission, 118*(5), 821–839.

Gilbert, R., Cohen, J. A., Pardo, S., Basu, A., & Fischman, D. A. (1999). Identification of the A-band localization domain of myosin binding proteins C and H (MyBP-C, MyBP-H) in skeletal muscle. *Journal of Cell Science, 112*(Pt. 1), 69–79.

Gini, B., Lovato, L., Cianti, R., Cecotti, L., Marconi, S., Anghileri, E., et al. (2008). Novel autoantigens recognized by CSF IgG from Hashimoto's encephalitis revealed by a proteomic approach. *Journal of Neuroimmunology, 196*(1–2), 153–158.

Giron, P., Dayon, L., Turck, N., Hoogland, C., & Sanchez, J. C. (2011). Quantitative analysis of human cerebrospinal fluid proteins using a combination of cysteine tagging and amine-reactive isobaric labeling. *Journal of Proteome Research, 10*(1), 249–258.

Glass, C. K., Saijo, K., Winner, B., Marchetto, M. C., & Gage, F. H. (2010). Mechanisms underlying inflammation in neurodegeneration. *Cell, 140*(6), 918–934.

Gonzalez de Aguilar, J. L., Niederhauser-Wiederkehr, C., Halter, B., De Tapia, M., Di Scala, F., Demougin, P., et al. (2008). Gene profiling of skeletal muscle in an amyotrophic lateral sclerosis mouse model. *Physiological Genomics, 32*(2), 207–218.

Gotz, J., & Ittner, L. M. (2008). Animal models of Alzheimer's disease and frontotemporal dementia. *Nature Reviews. Neuroscience, 9*(7), 532–544.

Grimm, S., Hoehn, A., Davies, K. J., & Grune, T. (2011). Protein oxidative modifications in the ageing brain: Consequence for the onset of neurodegenerative disease. *Free Radical Research, 45*(1), 73–88.

Harris, Z. L., Klomp, L. W., & Gitlin, J. D. (1998). Aceruloplasminemia: An inherited neurodegenerative disease with impairment of iron homeostasis. *The American Journal of Clinical Nutrition*, *67*(Suppl. 5), 972S–977S.

Hasegawa, M., Morishima-Kawashima, M., Takio, K., Suzuki, M., Titani, K., & Ihara, Y. (1992). Protein sequence and mass spectrometric analyses of tau in the Alzheimer's disease brain. *The Journal of Biological Chemistry*, *267*(24), 17047–17054.

Hellman, N. E., & Gitlin, J. D. (2002). Ceruloplasmin metabolism and function. *Annual Review of Nutrition*, *22*, 439–458.

Hosono, Y., Yamaguchi, T., Mizutani, E., Yanagisawa, K., Arima, C., Tomida, S., et al. (2011). MYBPH, a transcriptional target of TTF-1, inhibits ROCK1, and reduces cell motility and metastasis. *The EMBO Journal*, *31*(2), 481–493.

Hu, Y., Hosseini, A., Kauwe, J. S., Gross, J., Cairns, N. J., Goate, A. M., et al. (2007). Identification and validation of novel CSF biomarkers for early stages of Alzheimer's disease. *Proteomics. Clinical Applications*, *1*(11), 1373–1384.

Hu, Y., Malone, J. P., Fagan, A. M., Townsend, R. R., & Holtzman, D. M. (2005). Comparative proteomic analysis of intra- and interindividual variation in human cerebrospinal fluid. *Molecular & Cellular Proteomics: MCP*, *4*(12), 2000–2009.

Huang, J. T., McKenna, T., Hughes, C., Leweke, F. M., Schwarz, E., & Bahn, S. (2007). CSF biomarker discovery using label-free nano-LC-MS based proteomic profiling: Technical aspects. *Journal of Separation Science*, *30*(2), 214–225.

Iannaccone, S., Cerami, C., Alessio, M., Garibotto, V., Panzacchi, A., Olivieri, S., et al. (2013). In vivo microglia activation in very early dementia with Lewy bodies, comparison with Parkinson's disease. *Parkinsonism & Related Disorders*, *19*(1), 47–52.

Jagoe, R. T., Lecker, S. H., Gomes, M., & Goldberg, A. L. (2002). Patterns of gene expression in atrophying skeletal muscles: Response to food deprivation. *FASEB Journal: Official Publication of the Federation of American Societies for Experimental Biology*, *16*(13), 1697–1712.

Klade, C. S., Voss, T., Krystek, E., Ahorn, H., Zatloukal, K., Pummer, K., et al. (2001). Identification of tumor antigens in renal cell carcinoma by serological proteome analysis. *Proteomics*, *1*(7), 890–898.

Kono, S. (2012). Aceruloplasminemia. *Current Drug Targets*, *13*(9), 1190–1199.

Korschenhausen, D. A., Hampel, H. J., Ackenheil, M., Penning, R., & Muller, N. (1996). Fibrin degradation products in post mortem brain tissue of schizophrenics: A possible marker for underlying inflammatory processes. *Schizophrenia Research*, *19*(2–3), 103–109.

Krzyzanowska, A., & Carro, E. (2012). Pathological alteration in the choroid plexus of Alzheimer's disease: Implication for new therapy approaches. *Frontiers in Pharmacology*, *3*, 75.

Kuncl, R. W., Bilak, M. M., Bilak, S. R., Corse, A. M., Royal, W., & Becerra, S. P. (2002). Pigment epithelium-derived factor is elevated in CSF of patients with amyotrophic lateral sclerosis. *Journal of Neurochemistry*, *81*(1), 178–184.

Leverenz, J. B., Umar, I., Wang, Q., Montine, T. J., McMillan, P. J., Tsuang, D. W., et al. (2007). Proteomic identification of novel proteins in cortical Lewy bodies. *Brain Pathology*, *17*(2), 139–145.

Liao, L., Cheng, D., Wang, J., Duong, D. M., Losik, T. G., Gearing, M., et al. (2004). Proteomic characterization of postmortem amyloid plaques isolated by laser capture microdissection. *The Journal of Biological Chemistry*, *279*(35), 37061–37068.

Lista, S., Zetterberg, H., Dubois, B., Blennow, K., & Hampel, H. (2014). Cerebrospinal fluid analysis in Alzheimer's disease: Technical issues and future developments. *Journal of Neurology*, *261*(6), 1234–1243.

Liu, Y., Qing, H., & Deng, Y. (2014). Biomarkers in Alzheimer's disease analysis by mass spectrometry-based proteomics. *International Journal of Molecular Sciences*, *15*(5), 7865–7882.

Lourenco, A. S., Baldeiras, I., Graos, M., & Duarte, C. B. (2011). Proteomics-based technologies in the discovery of biomarkers for multiple sclerosis in the cerebrospinal fluid. *Current Molecular Medicine, 11*(4), 326–349.

Lovato, L., Cianti, R., Gini, B., Marconi, S., Bianchi, L., Armini, A., et al. (2008). Transketolase and 2′,3′-cyclic-nucleotide 3′-phosphodiesterase type I isoforms are specifically recognized by IgG autoantibodies in multiple sclerosis patients. *Molecular & Cellular Proteomics: MCP*, 7(12), 2337–2349.

Maik-Rachline, G., Shaltiel, S., & Seger, R. (2005). Extracellular phosphorylation converts pigment epithelium-derived factor from a neurotrophic to an antiangiogenic factor. *Blood, 105*(2), 670–678.

Menon, K. N., Steer, D. L., Short, M., Petratos, S., Smith, I., & Bernard, C. C. (2011). A novel unbiased proteomic approach to detect the reactivity of cerebrospinal fluid in neurological diseases. *Molecular & Cellular Proteomics: MCP, 10*(6), M110.00042.

Mila, S., Albo, A. G., Corpillo, D., Giraudo, S., Zibetti, M., Bucci, E. M., et al. (2009). Lymphocyte proteomics of Parkinson's disease patients reveals cytoskeletal protein dysregulation and oxidative stress. *Biomarkers in Medicine, 3*(2), 117–128.

Mittal, B., Doroudchi, M. M., Jeong, S. Y., Patel, B. N., & David, S. (2003). Expression of a membrane-bound form of the ferroxidase ceruloplasmin by leptomeningeal cells. *Glia, 41*(4), 337–346.

Mor, F., Izak, M., & Cohen, I. R. (2005). Identification of aldolase as a target antigen in Alzheimer's disease. *Journal of Immunology, 175*(5), 3439–3445.

Morell, A. G., Gregoriadis, G., Scheinberg, I. H., Hickman, J., & Ashwell, G. (1971). The role of sialic acid in determining the survival of glycoproteins in the circulation. *The Journal of Biological Chemistry, 246*(5), 1461–1467.

Morell, A. G., Irvine, R. A., Sternlieb, I., Scheinberg, I. H., & Ashwell, G. (1968). Physical and chemical studies on ceruloplasmin. V. Metabolic studies on sialic acid-free ceruloplasmin in vivo. *The Journal of Biological Chemistry, 243*(1), 155–159.

Musaro, A. (2010). State of the art and the dark side of amyotrophic lateral sclerosis. *World Journal of Biological Chemistry, 1*(5), 62–68.

Musci, G., Bonaccorsi di Patti, M. C., Fagiolo, U., & Calabrese, L. (1993). Age-related changes in human ceruloplasmin. Evidence for oxidative modifications. *The Journal of Biological Chemistry, 268*(18), 13388–13395.

Nakamura, T., Tu, S., Akhtar, M. W., Sunico, C. R., Okamoto, S., & Lipton, S. A. (2013). Aberrant protein s-nitrosylation in neurodegenerative diseases. *Neuron, 78*(4), 596–614.

Niebroj-Dobosz, I., Janik, P., Jamrozik, Z., & Kwiecinski, H. (1999). Immunochemical quantification of glycoconjugates in serum and cerebrospinal fluid of amyotrophic lateral sclerosis patients. *European Journal of Neurology, 6*(3), 335–340.

Nomura, T., Yabe, T., Mochizuki, H., Reiser, J., Becerra, S. P., & Schwartz, J. P. (2001). Survival effects of pigment epithelium-derived factor expressed by a lentiviral vector in rat cerebellar granule cells. *Developmental Neuroscience, 23*(2), 145–152.

Norman, H., Nordquist, J., Andersson, P., Ansved, T., Tang, X., Dworkin, B., et al. (2006). Impact of post-synaptic block of neuromuscular transmission, muscle unloading and mechanical ventilation on skeletal muscle protein and mRNA expression. *Pflügers Archiv, 453*(1), 53–66.

Nowacki, P., Stankowska-Chomicz, A., & Borodin, M. (1990). Fibrinogen degradation products in serum and cerebrospinal fluid in cerebrovascular diseases. *Neuropatologia Polska, 28*(3–4), 257–260.

Okamoto, S., Takahashi, H., Minoura, K., Fujita, N., & Yoshimura, M. (1991). FDP levels in the cerebrospinal fluid are elevated in patients with meningitis. *Rinsho Byori, 39*(6), 651–655.

Olivieri, S., Conti, A., Iannaccone, S., Cannistraci, C. V., Campanella, A., Barbariga, M., et al. (2011). Ceruloplasmin oxidation, a feature of Parkinson's disease CSF, inhibits ferroxidase activity and promotes cellular iron retention. *The Journal of Neuroscience*, *31*(50), 18568–18577.

Olson, M. F. (2008). Applications for ROCK kinase inhibition. *Current Opinion in Cell Biology*, *20*(2), 242–248.

Patel, B. N., Dunn, R. J., Jeong, S. Y., Zhu, Q., Julien, J. P., & David, S. (2002). Ceruloplasmin regulates iron levels in the CNS and prevents free radical injury. *The Journal of Neuroscience*, *22*(15), 6578–6586.

Perez-Gracia, E., Blanco, R., Carmona, M., Carro, E., & Ferrer, I. (2009). Oxidative stress damage and oxidative stress responses in the choroid plexus in Alzheimer's disease. *Acta Neuropathologica*, *118*(4), 497–504.

Perrin, R. J., Payton, J. E., Malone, J. P., Gilmore, P., Davis, A. E., Xiong, C., et al. (2013). Quantitative label-free proteomics for discovery of biomarkers in cerebrospinal fluid: Assessment of technical and inter-individual variation. *PLoS One*, *8*(5), e64314.

Petersen, S. V., Valnickova, Z., & Enghild, J. J. (2003). Pigment-epithelium-derived factor (PEDF) occurs at a physiologically relevant concentration in human blood: Purification and characterization. *The Biochemical Journal*, *374*(Pt. 1), 199–206.

Plebani, M., Sciacovelli, L., Aita, A., Padoan, A., & Chiozza, M. L. (2014). Quality indicators to detect pre-analytical errors in laboratory testing. *Clinica Chimica Acta; International Journal of Clinical Chemistry*, *432*, 44–48.

Poulsen, K., Bahl, J. M., Simonsen, A. H., Hasselbalch, S. G., & Heegaard, N. H. (2014). Distinct transthyretin oxidation isoform profile in spinal fluid from patients with Alzheimer's disease and mild cognitive impairment. *Clinical Proteomics*, *11*(1), 12.

Privitera, D., Corti, V., Alessio, M., Volonte, M. A., Lampasona, V., Comi, G., et al. (2013). Proteomic identification of aldolase A as an autoantibody target in patients with atypical movement disorders. *Neurological Sciences*, *34*(3), 313–320.

Ramirez-Castillejo, C., Sanchez-Sanchez, F., Andreu-Agullo, C., Ferron, S. R., Aroca-Aguilar, J. D., Sanchez, P., et al. (2006). Pigment epithelium-derived factor is a niche signal for neural stem cell renewal. *Nature Neuroscience*, *9*(3), 331–339.

Ravasi, T., Suzuki, H., Cannistraci, C. V., Katayama, S., Bajic, V. B., Tan, K., et al. (2010). An atlas of combinatorial transcriptional regulation in mouse and man. *Cell*, *140*(5), 744–752.

Reiber, H. (2001). Dynamics of brain-derived proteins in cerebrospinal fluid. *Clinica Chimica Acta; International Journal of Clinical Chemistry*, *310*(2), 173–186.

Rogowska-Wrzesinska, A., Le Bihan, M. C., Thaysen-Andersen, M., & Roepstorff, P. (2013). 2D gels still have a niche in proteomics. *Journal of Proteomics*, *88*, 4–13.

Sanger, T. D., Chen, D., Fehlings, D. L., Hallett, M., Lang, A. E., Mink, J. W., et al. (2010). Definition and classification of hyperkinetic movements in childhood. *Movement Disorders*, *25*(11), 1538–1549.

Schonberger, S. J., Edgar, P. F., Kydd, R., Faull, R. L., & Cooper, G. J. (2001). Proteomic analysis of the brain in Alzheimer's disease: Molecular phenotype of a complex disease process. *Proteomics*, *1*(12), 1519–1528.

Schutzer, S. E., Liu, T., Natelson, B. H., Angel, T. E., Schepmoes, A. A., Purvine, S. O., et al. (2010). Establishing the proteome of normal human cerebrospinal fluid. *PLoS One*, *5*(6), e10980.

Sedlak, E., Zoldak, G., & Wittung-Stafshede, P. (2008). Role of copper in thermal stability of human ceruloplasmin. *Biophysical Journal*, *94*(4), 1384–1391.

Seliger, B., & Kellner, R. (2002). Design of proteome-based studies in combination with serology for the identification of biomarkers and novel targets. *Proteomics*, *2*(12), 1641–1651.

Serot, J. M., Zmudka, J., & Jouanny, P. (2012). A possible role for CSF turnover and choroid plexus in the pathogenesis of late onset Alzheimer's disease. *Journal of Alzheimer's Disease: JAD*, *30*(1), 17–26.

Shimizu, T., Matsuoka, Y., & Shirasawa, T. (2005). Biological significance of isoaspartate and its repair system. *Biological & Pharmaceutical Bulletin, 28*(9), 1590–1596.

Shimizu, T., Watanabe, A., Ogawara, M., Mori, H., & Shirasawa, T. (2000). Isoaspartate formation and neurodegeneration in Alzheimer's disease. *Archives of Biochemistry and Biophysics, 381*(2), 225–234.

Shiozaki, A., Tsuji, T., Kohno, R., Kawamata, J., Uemura, K., Teraoka, H., et al. (2004). Proteome analysis of brain proteins in Alzheimer's disease: Subproteomics following sequentially extracted protein preparation. *Journal of Alzheimer's Disease: JAD, 6*(3), 257–268.

Smolinska, A., Blanchet, L., Buydens, L. M., & Wijmenga, S. S. (2012). NMR and pattern recognition methods in metabolomics: From data acquisition to biomarker discovery: A review. *Analytica Chimica Acta, 750*, 82–97.

Stoop, M. P., Coulier, L., Rosenling, T., Shi, S., Smolinska, A. M., Buydens, L., et al. (2010). Quantitative proteomics and metabolomics analysis of normal human cerebrospinal fluid samples. *Molecular & Cellular Proteomics: MCP, 9*(9), 2063–2075.

Stoop, M. P., Singh, V., Dekker, L. J., Titulaer, M. K., Stingl, C., Burgers, P. C., et al. (2010). Proteomics comparison of cerebrospinal fluid of relapsing remitting and primary progressive multiple sclerosis. *PLoS One, 5*(8), e12442.

Sultana, R., Reed, T., Perluigi, M., Coccia, R., Pierce, W. M., & Butterfield, D. A. (2007). Proteomic identification of nitrated brain proteins in amnestic mild cognitive impairment: A regional study. *Journal of Cellular and Molecular Medicine, 11*(4), 839–851.

Teunissen, C. E., Petzold, A., Bennett, J. L., Berven, F. S., Brundin, L., Comabella, M., et al. (2009). A consensus protocol for the standardization of cerebrospinal fluid collection and biobanking. *Neurology, 73*(22), 1914–1922.

Texel, S. J., Xu, X., & Harris, Z. L. (2008). Ceruloplasmin in neurodegenerative diseases. *Biochemical Society Transactions, 36*(Pt. 6), 1277–1281.

Thouvenot, E., Lafon-Cazal, M., Demettre, E., Jouin, P., Bockaert, J., & Marin, P. (2006). The proteomic analysis of mouse choroid plexus secretome reveals a high protein secretion capacity of choroidal epithelial cells. *Proteomics, 6*(22), 5941–5952.

Tohgi, H., Abe, T., Yamazaki, K., Murata, T., Ishizaki, E., & Isobe, C. (1999). Alterations of 3-nitrotyrosine concentration in the cerebrospinal fluid during aging and in patients with Alzheimer's disease. *Neuroscience Letters, 269*(1), 52–54.

Tombran-Tink, J., & Barnstable, C. J. (2003). PEDF: A multifaceted neurotrophic factor. *Nature Reviews. Neuroscience, 4*(8), 628–636.

Tovi, D., & Nilsson, I. M. (1972). Increased fibrinolytic activity and fibrin degradation products after experimental intracerebral haemorrhage. *Acta Neurologica Scandinavica, 48*(4), 403–415.

van de Warrenburg, B. P., Church, A. J., Martino, D., Candler, P. M., Bhatia, K. P., Giovannoni, G., et al. (2008). Antineuronal antibodies in Parkinson's disease. *Movement Disorders, 23*(7), 958–963.

van Gool, A. J., & Hendrickson, R. C. (2012). The proteomic toolbox for studying cerebrospinal fluid. *Expert Review of Proteomics, 9*(2), 165–179.

Wakabayashi, K., Tanji, K., Mori, F., & Takahashi, H. (2007). The Lewy body in Parkinson's disease: Molecules implicated in the formation and degradation of alpha-synuclein aggregates. *Neuropathology: Official Journal of the Japanese Society of Neuropathology, 27*(5), 494–506.

Waller, L. N., Shores, K., & Knapp, D. R. (2008). Shotgun proteomic analysis of cerebrospinal fluid using off-gel electrophoresis as the first-dimension separation. *Journal of Proteome Research, 7*(10), 4577–4584.

Wang, Q., Woltjer, R. L., Cimino, P. J., Pan, C., Montine, K. S., Zhang, J., et al. (2005). Proteomic analysis of neurofibrillary tangles in Alzheimer disease identifies GAPDH as a detergent-insoluble paired helical filament tau binding protein. *FASEB Journal: Official Publication of the Federation of American Societies for Experimental Biology, 19*(7), 869–871.

Ward, R. J., Zucca, F. A., Duyn, J. H., Crichton, R. R., & Zecca, L. (2014). The role of iron in brain ageing and neurodegenerative disorders. *Lancet Neurology, 13*(10), 1045–1060.

Watanabe, A., Takio, K., & Ihara, Y. (1999). Deamidation and isoaspartate formation in smeared tau in paired helical filaments. Unusual properties of the microtubule-binding domain of tau. *The Journal of Biological Chemistry, 274*(11), 7368–7378.

Weintraub, S. J., & Deverman, B. E. (2007). Chronoregulation by asparagine deamidation. *Science's STKE, 2007*(409), re7.

Wildenauer, D. B., Korschenhausen, D., Hoechtlen, W., Ackenheil, M., Kehl, M., & Lottspeich, F. (1991). Analysis of cerebrospinal fluid from patients with psychiatric and neurological disorders by two-dimensional electrophoresis: Identification of disease-associated polypeptides as fibrin fragments. *Electrophoresis, 12*(7–8), 487–492.

Wong, M., & Martin, L. J. (2010). Skeletal muscle-restricted expression of human SOD1 causes motor neuron degeneration in transgenic mice. *Human Molecular Genetics, 19*(11), 2284–2302.

Yasui, N., Mori, T., Morito, D., Matsushita, O., Kourai, H., Nagata, K., et al. (2003). Dual-site recognition of different extracellular matrix components by anti-angiogenic/neurotrophic serpin, PEDF. *Biochemistry, 42*(11), 3160–3167.

Yuan, X., & Desiderio, D. M. (2003). Proteomics analysis of phosphotyrosyl-proteins in human lumbar cerebrospinal fluid. *Journal of Proteome Research, 2*(5), 476–487.

Yuan, X., & Desiderio, D. M. (2005a). Proteomics analysis of human cerebrospinal fluid. *Journal of Chromatography. B, Analytical Technologies in the Biomedical and Life Sciences, 815*(1–2), 179–189.

Yuan, X., & Desiderio, D. M. (2005b). Proteomics analysis of prefractionated human lumbar cerebrospinal fluid. *Proteomics, 5*(2), 541–550.

Zhang, J. (2007). Proteomics of human cerebrospinal fluid—The good, the bad, and the ugly. *Proteomics Clinical Applications, 1*(8), 805–819.

Zhang, J., Goodlett, D. R., Peskind, E. R., Quinn, J. F., Zhou, Y., Wang, Q., et al. (2005). Quantitative proteomic analysis of age-related changes in human cerebrospinal fluid. *Neurobiology of Aging, 26*(2), 207–227.

Zhang, J., Goodlett, D. R., Quinn, J. F., Peskind, E., Kaye, J. A., Zhou, Y., et al. (2005). Quantitative proteomics of cerebrospinal fluid from patients with Alzheimer disease. *Journal of Alzheimer's Disease: JAD*, 7(2), 125–133, discussion 173–180.

Zhang, J., Keene, C. D., Pan, C., Montine, K. S., & Montine, T. J. (2008). Proteomics of human neurodegenerative diseases. *Journal of Neuropathology and Experimental Neurology, 67*(10), 923–932.

INDEX

Note: Page numbers followed by "*f*" indicate figures.

A

Advanced glycation end-products (AGEs), 100
Age-related macular degeneration (AMD), 60–61, 64
Alzheimer's disease (AD)
- animal studies, 15–16
- brain aging and longevity, 13
- brain transcriptome studies, 8–10
- copy number variation studies, 7
- early-onset, 11
- epigenomics studies, 11–13
- functional genomics studies, 8
- genome-wide association studies, 3–6
- iPSC technology, 16
- key factors and events, 2–3, 2*f*
- multiomics information, 17–18
- pathophysiology, 63
- peripheral biomarkers, 14–15
- proteomics, 44–47
- and relevant diseases, 13–14
- WES and WGS studies, 7–8

Alzheimer's disease biomarkers
- brain tissue, 63–64
- cerebrospinal fluid, 64
- plasma, 64–68

Amyloid, 88–89
Amyotrophic lateral sclerosis (ALS)
- nonsialylated ceruloplasmin, 138–140
- pathophysiology, 78
- proteomics, 46–47
- skeletal muscle involvement, 135–138

Amyotrophic lateral sclerosis biomarkers
- biofluids, 80
- CSF, 79
- plasma, 79–80

Apolipoprotein A-I (ApoA-I), 45
Atypical parkinsonian disorders, 47

B

Biofluids, 80
Brain aging, and longevity, 13
Brain amyloidosis, 89
Brain proteome, 94–96
Brain transcriptome, 8–10

C

Cancer, immunoproteomics, 124–125
Candidate biomarkers, 118–119
Carbamylation, 99
Carbonylation, 99–100
Cerebrospinal fluid (CSF)
- AD biomarkers, 64
- ALS biomarkers, 79–80
- prion diseases, 77–78

Chemical proteomics, 29–30
Clinical proteomics, 29–30, 118
Cognitive impairment
- cortical and subcortical dementias, 101–102
- ion channel proteins Na(+)K(+)-ATPase and dementia, 102–104
- oligomeric intermediates, 101
- protein misfolding, 100–101
- synaptic proteins and dementia, 104–106

Copy number variation (CNV), 7
Cross-linked fibrin degradation products (FDPs), 129–130

D

Degenerative protein modifications (DPMs)
- advanced glycation end-products, 100
- carbamylation, 99
- carbonylation, 99–100
- mass spectrometry-based proteomics, 92–94
- oxidation and nitration, 96–98
- racemization and isomerization, 98–99

Differential quantitative proteomics, 119–121
Diglycine (DiGly) capture proteomics, 29–30
DPMs. *See* Degenerative protein modifications (DPMs)

E

Early-onset AD, 11
Electrospray ionization (ESI), 62
Epigenomics, 11–13
Essential neuroproteomics
  biocomputational analysis, 126–127
  differential quantitative proteomics, 119–121
  immune proteomics, 123–126
  PTMs, 127–144
  sample tissues, 122–123

F

Functional genomics, 8
Functional proteomics, 29–30, 96

G

Genome-wide association studies (GWAS)
  primary phenotype of AD, 3–5
  secondary phenotype of AD, 5–6

H

Human brain homogenate, 120–121
Huntingtin (HTT) protein, 28
Huntington's disease (HD)
  htt protein, 74–76
  mitochondrial dysfunction, 72–73
  oxidatively modified proteins, 73–74
  pathophysiology, 71–72
  proteomics, 44–47
Hyperkinetic movement disorders (HMDs), 125–126

I

Immune proteomics, 123–126
Ion channel proteins Na(+)K(+)-ATPase, 102–104
iPSC technology, 16
Iso-aspartic acid, 90–91
Iso-glutamic acid, 90–91
Isomerization, 98–99

K

Kinomics, 29–30

L

Liquid chromatography (LC)-based approach, 119–120

M

Mass spectrometry (MS), 29–30, 62
  proteomics, 92–94
Matrix-assisted laser desorption ionization (MALDI), 62
Matrix metalloproteinase-9 (MMP-9), 79–80
Multidimensional protein identification technology (MudPIT), 31
Mutant HTT (mHTT), 28
Myosin-binding protein H (MyBP-H), 135–138

N

Neurodegeneration
  application of proteomics, 34–47
  cellular mechanisms, 27–28
  combinatorial genomic, metabolomics, and proteomic analysis, 48–49
  genetic causes and risk factors, 26
  proteomic protocols, 30–33
  signaling pathways, 47–48
  specifics of proteomic protocols, 33–34
  various diseases, 26–27, 27*f*
Neurofibrillary tangles (NFTs), 2
Nitration, 96–98
Nonenzymatic protein modifications, 89–92
Nonhypothesis-driven shotgun proteomics approach, 121
Nonsialylated ceruloplasmin, 138–140

O

Oxidation, 96–98
Oxidatively modified proteins, 73–74

P

Parkinson's disease (PD)
  pathophysiology, 68–69
  proteomics, 44–47
PD biomarkers
  brain tissue, 70–71
  cerebrospinal Fluid, 69–70
  plasma, 70

Peripheral biomarkers, 14–15
Pigment epithelium-derived factor (PEDF), 132
PLD3 gene, 7–8
Posttranslational modifications (PTMs)
  ALS patients, 138–140
  ceruloplasmin deamidation, 142–144
  ceruloplasmin oxidation, 140–142
  2DE protein separation, 127–128
  differentially expressed myosin-binding protein H, 135–138
  mild cognitive impairment patients, 128
  oxidative modifications, 128–129
  peripheral neuropathies, 132–135
  protein isoforms, 128
  traumatic brain injury patients, 129–131
Prion biomarkers
  brain, 76–77
  CSF, 77–78
Prion diseases, 76
Profiling proteomics, 96
Protein dysfunctions
  cognitive impairment, 100–106
  nonenzymatic protein modifications, 89–92
Protein–protein interaction (PPI) network approach, 9
Protein tyrosine nitration (PTN), 98
Proteomics
  AD biomarkers, 63–68
  ALS biomarkers, 79–80
  Alzheimer's disease, 44–47, 63
  amyotrophic lateral sclerosis, 46–47, 78
  animal models, 34–42
  atypical parkinsonian disorders, 47
  chemical, 29–30
  clinical, 29–30
  common protocols, 30–33
  CSF, 60–61
  diglycine capture, 29–30
  drug development, 60
  functional, 29–30
  human cells and tissue, 42–44
  Huntington's disease, 44–47, 71–76
  mass spectrometry, 62
  Parkinson's disease, 44–47, 68–69
  PD biomarkers, 69–71
  prion biomarkers, 76–78
  prion diseases, 76
  redox, 29–30
  specifics, 33–34
  two-dimensional gel electrophoresis, 61

R

Racemization, 98–99
Redox proteomics, 29–30

S

Schizophrenia (SZ), 13–14
Secretome, 123
Serological proteome analysis (SERPA), 124–125
Stable isotype dilution assay, 33
Superoxide dismutase 1 (SOD1), 28
Synaptic proteins, 104–106

T

Tetragonolobus purpureas agglutinin, 132–135
Transmissible spongiform encephalopathies (TSEs). *See* Prion diseases
Traumatic brain injury (TBI) patients, 129–131
TREM2 gene, 7
Two-dimensional electrophoresis (2DE), 119–120
Two-dimensional gel electrophoresis, 61

U

Ubiquitin-specific protease (USP9X), 34–42

V

Vascular dementia (VaD), 94–95

W

Whole-exome sequencing (WES), 7–8
Whole-genome sequencing (WGS), 7–8

CONTENTS OF RECENT VOLUMES

Volume 37

Section I: Selectionist Ideas and Neurobiology

Selectionist and Instructionist Ideas in Neuroscience
Olaf Sporns

Population Thinking and Neuronal Selection: Metaphors or Concepts?
Ernst Mayr

Selection and the Origin of Information
Manfred Eigen

Section II: Development and Neuronal Populations

Morphoregulatory Molecules and Selectional Dynamics during Development
Kathryn L. Crossin

Exploration and Selection in the Early Acquisition of Skill
Esther Thelen and Daniela Corbetta

Population Activity in the Control of Movement
Apostolos P. Georgopoulos

Section III: Functional Segregation and Integration in the Brain

Reentry and the Problem of Cortical Integration
Giulio Tononi

Coherence as an Organizing Principle of Cortical Functions
Wolf Singerl

Temporal Mechanisms in Perception
Ernst Pöppel

Section IV: Memory and Models

Selection versus Instruction: Use of Computer Models to Compare Brain Theories
George N. Reeke, Jr.

Memory and Forgetting: Long-Term and Gradual Changes in Memory Storage
Larry R. Squire

Implicit Knowledge: New Perspectives on Unconscious Processes
Daniel L. Schacter

Section V: Psychophysics, Psychoanalysis, and Neuropsychology

Phantom Limbs, Neglect Syndromes, Repressed Memories, and Freudian Psychology
V. S. Ramachandran

Neural Darwinism and a Conceptual Crisis in Psychoanalysis
Arnold H. Modell

A New Vision of the Mind
Oliver Sacks

INDEX

Volume 38

Regulation of $GABA_A$ Receptor Function and Gene Expression in the Central Nervous System
A. Leslie Morrow

Genetics and the Organization of the Basal Ganglia
Robert Hitzemann, Yeang Olan, Stephen Kanes, Katherine Dains, and Barbara Hitzemann

Structure and Pharmacology of Vertebrate $GABA_A$ Receptor Subtypes
Paul J. Whiting, Ruth M. McKernan, and Keith A. Wafford

Neurotransmitter Transporters: Molecular Biology, Function, and Regulation
Beth Borowsky and Beth J. Hoffman

Presynaptic Excitability
Meyer B. Jackson

Monoamine Neurotransmitters in Invertebrates and Vertebrates: An Examination of the Diverse Enzymatic Pathways Utilized to Synthesize and Inactivate Biogenic Amines
B. D. Sloley and A. V. Juorio

Neurotransmitter Systems in Schizophrenia
Gavin P. Reynolds

Physiology of Bergmann Glial Cells
Thomas Müller and Helmut Kettenmann

INDEX

Volume 39

Modulation of Amino Acid-Gated Ion Channels by Protein Phosphorylation
Stephen J. Moss and Trevor G. Smart

Use-Dependent Regulation of $GABA_A$ Receptors
Eugene M. Barnes, Jr.

Synaptic Transmission and Modulation in the Neostriatum
David M. Lovinger and Elizabeth Tyler

The Cytoskeleton and Neurotransmitter Receptors
Valerie J. Whatley and R. Adron Harris

Endogenous Opioid Regulation of Hippocampal Function
Michele L. Simmons and Charles Chavkin

Molecular Neurobiology of the Cannabinoid Receptor
Mary E. Abood and Billy R. Martin

Genetic Models in the Study of Anesthetic Drug Action
Victoria J. Simpson and Thomas E. Johnson

Neurochemical Bases of Locomotion and Ethanol Stimulant Effects
Tamara J. Phillips and Elaine H. Shen

Effects of Ethanol on Ion Channels
Fulton T. Crews, A. Leslie Morrow, Hugh Criswell, and George Breese

INDEX

Volume 40

Mechanisms of Nerve Cell Death: Apoptosis or Necrosis after Cerebral Ischemia
R. M. E. Chalmers-Redman, A. D. Fraser, W. Y. H. Ju, J. Wadia, N. A. Tatton, and W. G. Tatton

Changes in Ionic Fluxes during Cerebral Ischemia
Tibor Kristian and Bo K. Siesjo

Techniques for Examining Neuroprotective Drugs *in Vitro*
A. Richard Green and Alan J. Cross

Techniques for Examining Neuroprotective Drugs *in Vivo*
Mark P. Goldberg, Uta Strasser, and Laura L. Dugan

Calcium Antagonists: Their Role in Neuroprotection
A. Jacqueline Hunter

Sodium and Potassium Channel Modulators: Their Role in Neuroprotection
Tihomir P. Obrenovich

NMDA Antagonists: Their Role in Neuroprotection
Danial L. Small

Development of the NMDA Ion-Channel Blocker, Aptiganel Hydrochloride, as a Neuro-protective Agent for Acute CNS Injury
Robert N. McBurney

The Pharmacology of AMPA Antagonists and Their Role in Neuroprotection
Rammy Gill and David Lodge

GABA and Neuroprotection
Patrick D. Lyden

Adenosine and Neuroprotection
Bertil B. Fredholm

Interleukins and Cerebral Ischemia
Nancy J. Rothwell, Sarah A. Loddick, and Paul Stroemer

Nitrone-Based Free Radical Traps as Neuro-protective Agents in Cerebral Ischemia and Other Pathologies
Kenneth Hensley, John M. Carney, Charles A. Stewart, Tahera Tabatabaie, Quentin Pye, and Robert A. Floyd

Neurotoxic and Neuroprotective Roles of Nitric Oxide in Cerebral Ischemia
Turgay Dalkara and Michael A. Moskowitz

A Review of Earlier Clinical Studies on Neuro-protective Agents and Current Approaches
Nils-Gunnar Wahlgren

INDEX

Volume 41

Section I: Historical Overview

Rediscovery of an Early Concept
Jeremy D. Schmahmann

Section II: Anatomic Substrates

The Cerebrocerebellar System
Jeremy D. Schmahmann and Deepak N. Pandya

Cerebellar Output Channels
Frank A. Middleton and Peter L. Strick

Cerebellar-Hypothalamic Axis: Basic Circuits and Clinical Observations
Duane E. Haines, Espen Dietrichs, Gregory A. Mihailoff, and E. Frank McDonald

Section III. Physiological Observations

Amelioration of Aggression: Response to Selective Cerebellar Lesions in the Rhesus Monkey
Aaron J. Berman

Autonomic and Vasomotor Regulation
Donald J. Reis and Eugene V. Golanov

Associative Learning
Richard F. Thompson, Shaowen Bao, Lu Chen, Benjamin D. Cipriano, Jeffrey S. Grethe, Jeansok J. Kim, Judith K. Thompson, Jo Anne Tracy, Martha S. Weninger, and David J. Krupa

Visuospatial Abilities
Robert Lalonde

Spatial Event Processing
Marco Molinari, Laura Petrosini, and Liliana G. Grammaldo

Section IV: Functional Neuroimaging Studies

Linguistic Processing
Julie A. Fiez and Marcus E. Raichle

Sensory and Cognitive Functions
Lawrence M. Parsons and Peter T. Fox

Skill Learning
Julien Doyon

Section V: Clinical and Neuropsychological Observations

Executive Function and Motor Skill Learning
Mark Hallett and Jordon Grafman

Verbal Fluency and Agrammatism
Marco Molinari, Maria G. Leggio, and Maria C. Silveri

Classical Conditioning
Diana S. Woodruff-Pak

Early Infantile Autism
Margaret L. Bauman, Pauline A. Filipek, and Thomas L. Kemper

Olivopontocerebellar Atrophy and Friedreich's Ataxia: Neuropsychological Consequences of Bilateral versus Unilateral Cerebellar Lesions
Thérèse Botez-Marquard and Mihai I. Botez

Posterior Fossa Syndrome
Ian F. Pollack

Cerebellar Cognitive Affective Syndrome
Jeremy D. Schmahmann and Janet C. Sherman

Inherited Cerebellar Diseases
Claus W. Wallesch and Claudius Bartels

Neuropsychological Abnormalities in Cerebellar Syndromes—Fact or Fiction?
Irene Daum and Hermann Ackermann

Section VI: Theoretical Considerations

Cerebellar Microcomplexes
Masao Ito

Control of Sensory Data Acquisition
James M. Bower

Neural Representations of Moving Systems
Michael Paulin

How Fibers Subserve Computing Capabilities: Similarities between Brains and Machines
Henrietta C. Leiner and Alan L. Leiner

Cerebellar Timing Systems
Richard Ivry

Attention Coordination and Anticipatory Control
Natacha A. Akshoomoff, Eric Courchesne, and Jeanne Townsend

Context-Response Linkage
W. Thomas Thach

Duality of Cerebellar Motor and Cognitive Functions
James R. Bloedel and Vlastislav Bracha

Section VII: Future Directions

Therapeutic and Research Implications
Jeremy D. Schmahmann

Volume 42

Alzheimer Disease
Mark A. Smith

Neurobiology of Stroke
W. Dalton Dietrich

Free Radicals, Calcium, and the Synaptic Plasticity-Cell Death Continuum: Emerging Roles of the Trascription Factor NFκB
Mark P. Mattson

AP-I Transcription Factors: Short- and Long-Term Modulators of Gene Expression in the Brain
Keith Pennypacker

Ion Channels in Epilepsy
Istvan Mody

Posttranslational Regulation of Ionotropic Glutamate Receptors and Synaptic Plasticity
Xiaoning Bi, Steve Standley, and Michel Baudry

Heritable Mutations in the Glycine, $GABA_A$, and Nicotinic Acetylcholine Receptors Provide New Insights into the Ligand-Gated Ion Channel Receptor Superfamily
Behnaz Vafa and Peter R. Schofield

INDEX

Volume 43

Early Development of the *Drosophila* Neuromuscular Junction: A Model for Studying Neuronal Networks in Development
Akira Chiba

Development of Larval Body Wall Muscles
Michael Bate, Matthias Landgraf, and Mar Ruiz Gómez Bate

Development of Electrical Properties and Synaptic Transmission at the Embryonic Neuromuscular Junction
Kendal S. Broadie

Ultrastructural Correlates of Neuromuscular Junction Development
Mary B. Rheuben, Motojiro Yoshihara, and Yoshiaki Kidokoro

Assembly and Maturation of the *Drosophila* Larval Neuromuscular Junction
L. Sian Gramates and Vivian Budnik

Second Messenger Systems Underlying Plasticity at the Neuromuscular Junction
Frances Hannan and Yi Zhong

Mechanisms of Neurotransmitter Release
J. Troy Littleton, Leo Pallanck, and Barry Ganetzky

Vesicle Recycling at the *Drosophila* Neuromuscular Junction
Daniel T. Stimson and Mani Ramaswami

Ionic Currents in Larval Muscles of *Drosophila*
Satpal Singh and Chun-Fang Wu

Development of the Adult Neuromuscular System
Joyce J. Fernandes and Haig Keshishian

Controlling the Motor Neuron
James R. Trimarchi, Ping Jin, and Rodney K. Murphey

Volume 44

Human Ego-Motion Perception
A. V. van den Berg

Optic Flow and Eye Movements
M. Lappe and K.-P. Hoffman

The Role of MST Neurons during Ocular Tracking in 3D Space
K. Kawano, U. Inoue, A. Takemura, Y. Kodaka, and F. A. Miles

Visual Navigation in Flying Insects
M. V. Srinivasan and S.-W. Zhang

Neuronal Matched Filters for Optic Flow Processing in Flying Insects
H. G. Krapp

A Common Frame of Reference for the Analysis of Optic Flow and Vestibular Information
B. J. Frost and D. R. W. Wylie

Optic Flow and the Visual Guidance of Locomotion in the Cat
H. Sherk and G. A. Fowler

Stages of Self-Motion Processing in Primate Posterior Parietal Cortex
F. Bremmer, J.-R. Duhamel, S. B. Hamed, and W. Graf

Optic Flow Analysis for Self-Movement Perception
C. J. Duffy

Neural Mechanisms for Self-Motion Perception in Area MST
R. A. Andersen, K. V. Shenoy, J. A. Crowell, and D. C. Bradley

Computational Mechanisms for Optic Flow Analysis in Primate Cortex
M. Lappe

Human Cortical Areas Underlying the Perception of Optic Flow: Brain Imaging Studies
M. W. Greenlee

What Neurological Patients Tell Us about the Use of Optic Flow
L. M. Vaina and S. K. Rushton

INDEX

Volume 45

Mechanisms of Brain Plasticity: From Normal Brain Function to Pathology
Philip. A. Schwartzkroin

Brain Development and Generation of Brain Pathologies
Gregory L. Holmes and Bridget McCabe

Maturation of Channels and Receptors: Consequences for Excitability
David F. Owens and Arnold R. Kriegstein

Neuronal Activity and the Establishment of Normal and Epileptic Circuits during Brain Development
John W. Swann, Karen L. Smith, and Chong L. Lee

The Effects of Seizures of the Hippocampus of the Immature Brain
Ellen F. Sperber and Solomon L. Moshe

Abnormal Development and Catastrophic Epilepsies: The Clinical Picture and Relation to Neuroimaging
Harry T. Chugani and Diane C. Chugani

Cortical Reorganization and Seizure Generation in Dysplastic Cortex
G. Avanzini, R. Preafico, S. Franceschetti, G. Sancini, G. Battaglia, and V. Scaioli

Rasmussen's Syndrome with Particular Reference to Cerebral Plasticity: A Tribute to Frank Morrell
Fredrick Andermann and Yuonne Hart

Structural Reorganization of Hippocampal Networks Caused by Seizure Activity
Daniel H. Lowenstein

Epilepsy-Associated Plasticity in gamma-Amniobutyric Acid Receptor Expression, Function and Inhibitory Synaptic Properties
Douglas A. Coulter

Synaptic Plasticity and Secondary Epileptogenesis
Timothy J. Teyler, Steven L. Morgan, Rebecca N. Russell, and Brian L. Woodside

Synaptic Plasticity in Epileptogenesis: Cellular Mechanisms Underlying Long-Lasting Synaptic Modifications that Require New Gene Expression
Oswald Steward, Christopher S. Wallace, and Paul F. Worley

Cellular Correlates of Behavior
Emma R. Wood, Paul A. Dudchenko, and Howard Eichenbaum

Mechanisms of Neuronal Conditioning
David A. T. King, David J. Krupa, Michael R. Foy, and Richard F. Thompson

Plasticity in the Aging Central Nervous System
C. A. Barnes

Secondary Epileptogenesis, Kindling, and Intractable Epilepsy: A Reappraisal from the Perspective of Neuronal Plasticity
Thomas P. Sutula

Kindling and the Mirror Focus
Dan C. McIntyre and Michael O. Poulter

Partial Kindling and Behavioral Pathologies
Robert E. Adamec

The Mirror Focus and Secondary Epileptogenesis
B. J. Wilder

Hippocampal Lesions in Epilepsy: A Historical Review
Robert Naquet

Clinical Evidence for Secondary Epileptogensis
Hans O. Luders

Epilepsy as a Progressive (or Nonprogressive "Benign") Disorder
John A. Wada

Pathophysiological Aspects of Landau-Kleffner Syndrome: From the Active Epileptic Phase to Recovery
Marie-Noelle Metz-Lutz, Pierre Maquet, Annd De Saint Martin, Gabrielle Rudolf, Norma Wioland, Edouard Hirsch, and Chriatian Marescaux

Local Pathways of Seizure Propagation in Neocortex
Barry W. Connors, David J. Pinto, and Albert E. Telefeian

Multiple Subpial Transection: A Clinical Assessment
C. E. Polkey

The Legacy of Frank Morrell
Jerome Engel, Jr.

Volume 46

Neurosteroids: Beginning of the Story
Etienne E. Baulieu, P. Robel, and M. Schumacher

Biosynthesis of Neurosteroids and Regulation of Their Synthesis
Synthia H. Mellon and Hubert Vaudry

Neurosteroid 7-Hydroxylation Products in the Brain
Robert Morfin and Luboslav Stárka

Neurosteroid Analysis
Ahmed A. Alomary, Robert L. Fitzgerald, and Robert H. Purdy

Role of the Peripheral-Type Benzodiazepine Receptor in Adrenal and Brain Steroidogenesis
Rachel C. Brown and Vassilios Papadopoulos

Formation and Effects of Neuroactive Steroids in the Central and Peripheral Nervous System
Roberto Cosimo Melcangi, Valerio Magnaghi, Mariarita Galbiati, and Luciano Martini

Neurosteroid Modulation of Recombinant and Synaptic $GABA_A$ Receptors
Jeremy J. Lambert, Sarah C. Harney, Delia Belelli, and John A. Peters

$GABA_A$-Receptor Plasticity during Long-Term Exposure to and Withdrawal from Progesterone
Giovanni Biggio, Paolo Follesa, Enrico Sanna, Robert H. Purdy, and Alessandra Concas

Stress and Neuroactive Steroids
Maria Luisa Barbaccia, Mariangela Serra, Robert H. Purdy, and Giovanni Biggio

Neurosteroids in Learning and Memory Processes
Monique Vallée, Willy Mayo, George F. Koob, and Michel Le Moal

Neurosteroids and Behavior
Sharon R. Engel and Kathleen A. Grant

Ethanol and Neurosteroid Interactions in the Brain
A. Leslie Morrow, Margaret J. VanDoren, Rebekah Fleming, and Shannon Penland

Preclinical Development of Neurosteroids as Neuroprotective Agents for the Treatment of Neurodegenerative Diseases
Paul A. Lapchak and Dalia M. Araujo

Clinical Implications of Circulating Neurosteroids
Andrea R. Genazzani, Patrizia Monteleone, Massimo Stomati, Francesca Bernardi, Luigi Cobellis, Elena Casarosa, Michele Luisi, Stefano Luisi, and Felice Petraglia

Neuroactive Steroids and Central Nervous System Disorders
Mingde Wang, Torbjörn Bäckström, Inger Sundström, Göran Wahlström, Tommy Olsson, Di Zhu, Inga-Maj Johansson, Inger Björn, and Marie Bixo

Neuroactive Steroids in Neuropsychopharmacology
Rainer Rupprecht and Florian Holsboer

Current Perspectives on the Role of Neurosteroids in PMS and Depression
Lisa D. Griffin, Susan C. Conrad, and Synthia H. Mellon

Index

Volume 47

Introduction: Studying Gene Expression in Neural Tissues by *in Situ* Hybridization
W. Wisden and B. J. Morris

Part I: *In Situ* Hybridization with Radiolabelled Oligonucleotides
In Situ Hybridization with Oligonucleotide Probes
Wl. Wisden and B. J. Morris

Cryostat Sectioning of Brains
Victoria Revilla and Alison Jones

Processing Rodent Embryonic and Early Postnatal Tissue for *in Situ* Hybridization with Radiolabelled Oligonucleotides
David J. Laurie, Petra C. U. Schrotz, Hannah Monyer, and Ulla Amtmann

Processing of Retinal Tissue for *in Situ* Hybridization
Frank Müller

Processing the Spinal Cord for *in Situ* Hybridization with Radiolabelled Oligonucleotides
A. Berthele and T. R. Tölle

Processing Human Brain Tissue for *in Situ* Hybridization with Radiolabelled Oligonucleotides
Louise F. B. Nicholson

In Situ Hybridization of Astrocytes and Neurons Cultured *in Vitro*
L. A. Arizza-McNaughton, C. De Felipe, and S. P. Hunt

In Situ Hybridization on Organotypic Slice Cultures
A. Gerfin-Moser and H. Monyer

Quantitative Analysis of *in Situ* Hybridization Histochemistry
Andrew L. Gundlach and Ross D. O'Shea

Part II: Nonradioactive *in Situ* hybridization

Nonradioactive *in Situ* Hybridization Using Alkaline Phosphatase-Labelled Oligonucleotides
S. J. Augood, E. M. McGowan, B. R. Finsen, B. Heppelmann, and P. C. Emson

Combining Nonradioactive *in Situ* Hybridization with Immunohistological and Anatomical Techniques
Petra Wahle

Nonradioactive *in Situ* Hybridization: Simplified Procedures for Use in Whole Mounts of Mouse and Chick Embryos
Linda Ariza-McNaughton and Robb Krumlauf

INDEX

Volume 48

Assembly and Intracellular Trafficking of $GABA_A$ Receptors Eugene
Barnes

Subcellular Localization and Regulation of $GABA_A$ Receptors and Associated Proteins Bernhard Lüscher and Jean-Marc Fritschy D_1 Dopamine Receptors
Richard Mailman

Molecular Modeling of Ligand-Gated Ion Channels: Progress and Challenges
Ed Bertaccini and James R. Trudel

Alzheimer's Disease: Its Diagnosis and Pathogenesis
Jillian J. Kril and Glenda M. Halliday

DNA Arrays and Functional Genomics in Neurobiology
Christelle Thibault, Long Wang, Li Zhang, and Michael F. Miles

INDEX

Volume 49

What Is West Syndrome?
Olivier Dulac, Christine Soufflet, Catherine Chiron, and Anna Kaminski

The Relationship between encephalopathy and Abnormal Neuronal Activity in the Developing Brain
Frances E. Jensen

Hypotheses from Functional Neuroimaging Studies
Csaba Juhász, Harry T. Chugani, Ouo Muzik, and Diane C. Chugani

Infantile Spasms: Unique Sydrome or General Age-Dependent Manifestation of a Diffuse Encephalopathy?
M. A. Koehn and M. Duchowny

Histopathology of Brain Tissue from Patients with Infantile Spasms
Harry V. Vinters

Generators of Ictal and Interictal Electroencephalograms Associated with Infantile Spasms: Intracellular Studies of Cortical and Thalamic Neurons
M. Steriade and I. Timofeev

Cortical and Subcortical Generators of Normal and Abnormal Rhythmicity
David A. McCormick

Role of Subcortical Structures in the Pathogenesis of Infantile Spasms: What Are Possible Subcortical Mediators?
F. A. Lado and S. L. Moshé

What Must We Know to Develop Better Therapies?
Jean Aicardi

The Treatment of Infantile Spasms: An Evidence-Based Approach
Mark Mackay, Shelly Weiss, and O. Carter Snead III

ACTH Treatment of Infantile Spasms: Mechanisms of Its Effects in Modulation of Neuronal Excitability
K. L. Brunson, S. Avishai-Eliner, and T. Z. Baram

Neurosteroids and Infantile Spasms: The Deoxycorticosterone Hypothesis
Michael A. Rogawski and Doodipala S. Reddy

Are there Specific Anatomical and/or Transmitter Systems (Cortical or Subcortical) That Should Be Targeted?
Phillip C. Jobe

Medical versus Surgical Treatment: Which Treatment When
W. Donald Shields

Developmental Outcome with and without Successful Intervention
Rochelle Caplan, Prabha Siddarth, Gary Mathern, Harry Vinters, Susan Curtiss, Jennifer Levitt, Robert Asarnow, and W. Donald Shields

Infantile Spasms versus Myoclonus: Is There a Connection?
Michael R. Pranzatelli

Tuberous Sclerosis as an Underlying Basis for Infantile Spasm
Raymond S. Yeung

Brain Malformation, Epilepsy, and Infantile Spasms
M. Elizabeth Ross

Brain Maturational Aspects Relevant to Pathophysiology of Infantile Spasms
G. Auanzini, F. Panzica, and S. Franceschetti

Gene Expression Analysis as a Strategy to Understand the Molecular Pathogenesis of Infantile Spasms
Peter B. Crino

Infantile Spasms: Criteria for an Animal Model
Carl E. Stafstrom and Gregory L. Holmes

INDEX

Volume 50

Part I: Primary Mechanisms

How Does Glucose Generate Oxidative Stress In Peripheral Nerve?
Irina G. Obrosova

Glycation in Diabetic Neuropathy: Characteristics, Consequences, Causes, and Therapeutic Options
Paul J. Thornalley

Part II: Secondary Changes

Protein Kinase C Changes in Diabetes: Is the Concept Relevant to Neuropathy?
Joseph Eichberg

Are Mitogen-Activated Protein Kinases Glucose Transducers for Diabetic Neuropathies?
Tertia D. Purves and David R. Tomlinson

Neurofilaments in Diabetic Neuropathy
Paul Fernyhough and Robert E. Schmidt

Apoptosis in Diabetic Neuropathy
Aviva Tolkovsky

Nerve and Ganglion Blood Flow in Diabetes: An Appraisal
Douglas W. Zochodne

Part III: Manifestations

Potential Mechanisms of Neuropathic Pain in Diabetes
Nigel A. Calcutt

Electrophysiologic Measures of Diabetic Neuropathy: Mechanism and Meaning
Joseph C. Arezzo and Elena Zotova

Neuropathology and Pathogenesis of Diabetic Autonomic Neuropathy
Robert E. Schmidt

Role of the Schwann Cell in Diabetic Neuropathy
Luke Eckersley

Part IV: Potential Treatment

Polyol Pathway and Diabetic Peripheral Neuropathy
Peter J. Oates

Nerve Growth Factor for the Treatment of Diabetic Neuropathy: What Went Wrong, What Went Right, and What Does the Future Hold?
Stuart C. Apfel

Angiotensin-Converting Enzyme Inhibitors: Are there Credible Mechanisms for Beneficial Effects in Diabetic Neuropathy?
Rayaz A. Malik and David R. Tomlinson

Clinical Trials for Drugs Against Diabetic Neuropathy: Can We Combine Scientific Needs With Clinical Practicalities?
Dan Ziegler and Dieter Luft

INDEX

Volume 51

Energy Metabolism in the Brain
Leif Hertz and Gerald A. Dienel

The Cerebral Glucose-Fatty Acid Cycle: Evolutionary Roots, Regulation, and (Patho) physiological Importance
Kurt Heininger

Expression, Regulation, and Functional Role of Glucose Transporters (GLUTs) in Brain
Donard S. Dwyer, Susan J. Vannucci, and Ian A. Simpson

Insulin-Like Growth Factor-1 Promotes Neuronal Glucose Utilization During Brain Development and Repair Processes
Carolyn A. Bondy and Clara M. Cheng

CNS Sensing and Regulation of Peripheral Glucose Levels
Barry E. Levin, Ambrose A. Dunn-Meynell, and Vanessa H. Routh

Glucose Transporter Protein Syndromes
Darryl C. De Vivo, Dong Wang, Juan M. Pascual, and Yuan Yuan Ho

Glucose, Stress, and Hippocampal Neuronal Vulnerability
Lawrence P. Reagan

Glucose/Mitochondria in Neurological Conditions
John P. Blass

Energy Utilization in the Ischemic/Reperfused Brain
John W. Phillis and Michael H. O'Regan

Diabetes Mellitus and the Central Nervous System
Anthony L. McCall

Diabetes, the Brain, and Behavior: Is There a Biological Mechanism Underlying the Association between Diabetes and Depression?
A. M. Jacobson, J. A. Samson, K. Weinger, and C. M. Ryan

Schizophrenia and Diabetes
David C. Henderson and Elissa R. Ettinger

Psychoactive Drugs Affect Glucose Transport and the Regulation of Glucose Metabolism
Donard S. Dwyer, Timothy D. Ardizzone, and Ronald J. Bradley

INDEX

Volume 52

Neuroimmune Relationships in Perspective
Frank Hucklebridge and Angela Clow

Sympathetic Nervous System Interaction with the Immune System
Virginia M. Sanders and Adam P. Kohm

Mechanisms by Which Cytokines Signal the Brain
Adrian J. Dunn

Neuropeptides: Modulators of Immune Responses in Health and Disease
David S. Jessop

Brain–Immune Interactions in Sleep
Lisa Marshall and Jan Born

Neuroendocrinology of Autoimmunity
Michael Harbuz

Systemic Stress-Induced Th2 Shift and Its Clinical Implications
Ibia J. Elenkov

Neural Control of Salivary S-IgA Secretion
Gordon B. Proctor and Guy H. Carpenter

Stress and Secretory Immunity
Jos A. Bosch, Christopher Ring, Eco J. C. de Geus, Enno C. I. Veerman, and Arie V. Nieuw Amerongen

Cytokines and Depression
Angela Clow

Immunity and Schizophrenia: Autoimmunity, Cytokines, and Immune Responses
Fiona Gaughran

Cerebral Lateralization and the Immune System
Pierre J. Neveu

Behavioral Conditioning of the Immune System
Frank Hucklebridge

Psychological and Neuroendocrine Correlates of Disease Progression
Julie M. Turner-Cobb

The Role of Psychological Intervention in Modulating Aspects of Immune Function in Relation to Health and Well-Being
J. H. Gruzelier

INDEX

Volume 53

Section I: Mitochondrial Structure and Function

Mitochondrial DNA Structure and Function
Carlos T. Moraes, Sarika Srivastava, Ilias Kirkinezos, Jose Oca-Cossio, Corina van Waveren, Markus Woischnick, and Francisca Diaz

Oxidative Phosphorylation: Structure, Function, and Intermediary Metabolism
Simon J. R. Heales, Matthew E. Gegg, and John B. Clark

Import of Mitochondrial Proteins
Matthias F. Bauer, Sabine Hofmann, and Walter Neupert

Section II: Primary Respiratory Chain Disorders

Mitochondrial Disorders of the Nervous System: Clinical, Biochemical, and Molecular Genetic Features
Dominic Thyagarajan and Edward Byrne

Section III: Secondary Respiratory Chain Disorders

Friedreich's Ataxia
J. M. Cooper and J. L. Bradley

Wilson Disease
C. A. Davie and A. H. V. Schapira

Hereditary Spastic Paraplegia
Christopher J. McDermott and Pamela J. Shaw

Cytochrome c Oxidase Deficiency
Giacomo P. Comi, Sandra Strazzer, Sara Galbiati, and Nereo Bresolin

Section IV: Toxin Induced Mitochondrial Dysfunction

Toxin-Induced Mitochondrial Dysfunction
Susan E. Browne and M. Flint Beal

Section V: Neurodegenerative Disorders

Parkinson's Disease
L. V. P. Korlipara and A. H. V. Schapira

Huntington's Disease: The Mystery Unfolds?
Åsa Petersén and Patrik Brundin

Mitochondria in Alzheimer's Disease
Russell H. Swerdlow and Stephen J. Kish

Contributions of Mitochondrial Alterations, Resulting from Bad Genes and a Hostile Environment, to the Pathogenesis of Alzheimer's Disease
Mark P. Mattson

Mitochondria and Amyotrophic Lateral Sclerosis
Richard W. Orrell and Anthony H. V. Schapira

Section VI: Models of Mitochondrial Disease

Models of Mitochondrial Disease
Danae Liolitsa and Michael G. Hanna

Section VII: Defects of β Oxidation Including Carnitine Deficiency

Defects of β Oxidation Including Carnitine Deficiency
K. Bartlett and M. Pourfarzam

Section VIII: Mitochondrial Involvement in Aging

The Mitochondrial Theory of Aging: Involvement of Mitochondrial DNA Damage and Repair
Nadja C. de Souza-Pinto and Vilhelm A. Bohr

INDEX

Volume 54

Unique General Anesthetic Binding Sites Within Distinct Conformational States of the Nicotinic Acetylcholine Réceptor
Hugo R. Ariaas, William, R. Kem, James R. Truddell, and Michael P. Blanton

Signaling Molecules and Receptor Transduction Cascades That Regulate NMDA Receptor-Mediated Synaptic Transmission
Suhas. A. Kotecha and John F. MacDonald

Behavioral Measures of Alcohol Self-Administration and Intake Control: Rodent Models
Herman H. Samson and Cristine L. Czachowski

Dopaminergic Mouse Mutants: Investigating the Roles of the Different Dopamine Receptor Subtypes and the Dopamine Transporter
Shirlee Tan, Bettina Hermann, and Emiliana Borrelli

Drosophila melanogaster, A Genetic Model System for Alcohol Research
Douglas J. Guarnieri and Ulrike Heberlein

INDEX

Volume 55

Section I: Virsu Vectors For Use in the Nervous System

Non-Neurotropic Adenovirus: a Vector for Gene Transfer to the Brain and Gene Therapy of Neurological Disorders
P. R. Lowenstein, D. Suwelack, J. Hu, X. Yuan, M. Jimenez-Dalmaroni, S. Goverdhama, and M.G. Castro

Adeno-Associated Virus Vectors
E. Lehtonen and L. Tenenbaum

Problems in the Use of Herpes Simplex Virus as a Vector
L. T. Feldman

Lentiviral Vectors
J. Jakobsson, C. Ericson, N. Rosenquist, and C. Lundberg

Retroviral Vectors for Gene Delivery to Neural Precursor Cells
K. Kageyama, H. Hirata, and J. Hatakeyama

Section II: Gene Therapy with Virus Vectors for Specific Disease of the Nervous System

The Principles of Molecular Therapies for Glioblastoma
G. Karpati and J. Nalbatonglu

Oncolytic Herpes Simplex Virus
J. C. C. Hu and R. S. Coffin

Recombinant Retrovirus Vectors for Treatment of Brain Tumors
N. G. Rainov and C. M. Kramm

Adeno-Associated Viral Vectors for Parkinson's Disease
I. Muramatsu, L. Wang, K. Ikeguchi, K-i Fujimoto, T. Okada, H. Mizukami, Y. Hanazono, A. Kume, I. Nakano, and K. Ozawa

HSV Vectors for Parkinson's Disease
D. S. Latchman

Gene Therapy for Stroke
K. Abe and W. R. Zhang

Gene Therapy for Mucopolysaccharidosis
A. Bosch and J. M. Heard

INDEX

Volume 56

Behavioral Mechanisms and the Neurobiology of Conditioned Sexual Responding
Mark Krause

NMDA Receptors in Alcoholism
Paula L. Hoffman

Processing and Representation of Species-Specific Communication Calls in the Auditory System of Bats
George D. Pollak, Achim Klug, and Eric E. Bauer

Central Nervous System Control of Micturition
Gert Holstege and Leonora J. Mouton

The Structure and Physiology of the Rat Auditory System: An Overview
Manuel Malmierca

Neurobiology of Cat and Human Sexual Behavior
Gert Holstege and J. R. Georgiadis

INDEX

Volume 57

Cumulative Subject Index of Volumes 1–25

Volume 58

Cumulative Subject Index of Volumes 26–50

Volume 59

Loss of Spines and Neuropil
Liesl B. Jones

Schizophrenia as a Disorder of Neuroplasticity
Robert E. McCullumsmith, Sarah M. Clinton, and James H. Meador-Woodruff

The Synaptic Pathology of Schizophrenia: Is Aberrant Neurodevelopment and Plasticity to Blame?
Sharon L. Eastwood

Neurochemical Basis for an Epigenetic Vision of Synaptic Organization
E. Costa, D. R. Grayson, M. Veldic, and A. Guidotti

Muscarinic Receptors in Schizophrenia: Is There a Role for Synaptic Plasticity?
Thomas J. Raedler

Serotonin and Brain Development
Monsheel S. K. Sodhi and Elaine Sanders-Bush

Presynaptic Proteins and Schizophrenia
William G. Honer and Clint E. Young

Mitogen-Activated Protein Kinase Signaling
Svetlana V. Kyosseva

Postsynaptic Density Scaffolding Proteins at Excitatory Synapse and Disorders of Synaptic Plasticity: Implications for Human Behavior Pathologies
Andrea de Bartolomeis and Germano Fiore

Prostaglandin-Mediated Signaling in Schizophrenia
S. Smesny

Mitochondria, Synaptic Plasticity, and Schizophrenia
Dorit Ben-Shachar and Daphna Laifenfeld

Membrane Phospholipids and Cytokine Interaction in Schizophrenia
Jeffrey K. Yao and Daniel P. van Kammen

Neurotensin, Schizophrenia, and Antipsychotic Drug Action
Becky Kinkead and Charles B. Nemeroff

Schizophrenia, Vitamin D, and Brain Development
Alan Mackay-Sim, François FÉron, Darryl Eyles, Thomas Burne, and John McGrath

Possible Contributions of Myelin and Oligodendrocyte Dysfunction to Schizophrenia
Daniel G. Stewart and Kenneth L. Davis

Brain-Derived Neurotrophic Factor and the Plasticity of the Mesolimbic Dopamine Pathway
Oliver Guillin, Nathalie Griffon, Jorge Diaz, Bernard Le Foll, Erwan Bezard, Christian Gross, Chris Lammers, Holger Stark, Patrick Carroll, Jean-Charles Schwartz, and Pierre Sokoloff

S100B in Schizophrenic Psychosis
Matthias Rothermundt, Gerald Ponath, and Volker Arolt

Oct-6 Transcription Factor
Maria Ilia

NMDA Receptor Function, Neuroplasticity, and the Pathophysiology of Schizophrenia
Joseph T. Coyle and Guochuan Tsai

INDEX

Volume 60

Microarray Platforms: Introduction and Application to Neurobiology
Stanislav L. Karsten, Lili C. Kudo, and Daniel H. Geschwind

Experimental Design and Low-Level Analysis of Microarray Data
B. M. Bolstad, F. Collin, K. M. Simpson, R. A. Irizarry, and T. P. Speed

Brain Gene Expression: Genomics and Genetics
Elissa J. Chesler and Robert W. Williams

DNA Microarrays and Animal Models of Learning and Memory
Sebastiano Cavallaro

Microarray Analysis of Human Nervous System Gene Expression in Neurological Disease
Steven A. Greenberg

DNA Microarray Analysis of Postmortem Brain Tissue
Károly Mirnics, Pat Levitt, and David A. Lewis

INDEX

Volume 61

Section I: High-Throughput Technologies

Biomarker Discovery Using Molecular Profiling Approaches
Stephen J. Walker and Arron Xu

Proteomic Analysis of Mitochondrial Proteins
Mary F. Lopez, Simon Melov, Felicity Johnson, Nicole Nagulko, Eva Golenko, Scott Kuzdzal, Suzanne Ackloo, and Alvydas Mikulskis

Section II: Proteomic Applications

NMDA Receptors, Neural Pathways, and Protein Interaction Databases
Holger Husi

Dopamine Transporter Network and Pathways
Rajani Maiya and R. Dayne Mayfield

Proteomic Approaches in Drug Discovery and Development
Holly D. Soares, Stephen A. Williams, Peter J. Snyder, Feng Gao, Tom Stiger, Christian Rohlff, Athula Herath, Trey Sunderland, Karen Putnam, and W. Frost White

Section III: Informatics

Proteomic Informatics
Steven Russell, William Old, Katheryn Resing, and Lawrence Hunter

Section IV: Changes in the Proteome by Disease

Proteomics Analysis in Alzheimer's Disease: New Insights into Mechanisms of Neurodegeneration
D. Allan Butterfield and Debra Boyd-Kimball

Proteomics and Alcoholism
Frank A. Witzmann and Wendy N. Strother

Proteomics Studies of Traumatic Brain Injury
Kevin K. W. Wang, Andrew Ottens, William Haskins, Ming Cheng Liu, Firas Kobeissy, Nancy Denslow, SuShing Chen, and Ronald L. Hayes

Influence of Huntington's Disease on the Human and Mouse Proteome
Claus Zabel and Joachim Klose

Section V: Overview of the Neuroproteome

Proteomics—Application to the Brain
Katrin Marcus, Oliver Schmidt, Heike Schaefer, Michael Hamacher, AndrÅ van Hall, and Helmut E. Meyer

INDEX

Volume 62

$GABA_A$ Receptor Structure–Function Studies: A Reexamination in Light of New Acetylcholine Receptor Structures
Myles H. Akabas

Dopamine Mechanisms and Cocaine Reward
Aiko Ikegami and Christine L. Duvauchelle

Proteolytic Dysfunction in Neurodegenerative Disorders
Kevin St. P. McNaught

Neuroimaging Studies in Bipolar Children and Adolescents
Rene L. Olvera, David C. Glahn, Sheila C. Caetano, Steven R. Pliszka, and Jair C. Soares

Chemosensory G-Protein-Coupled Receptor Signaling in the Brain
Geoffrey E. Woodard

Disturbances of Emotion Regulation after Focal Brain Lesions
Antoine Bechara

The Use of *Caenorhabditis elegans* in Molecular Neuropharmacology
Jill C. Bettinger, Lucinda Carnell, Andrew G. Davies, and Steven L. McIntire

INDEX

Volume 63

Mapping Neuroreceptors at work: On the Definition and Interpretation of Binding Potentials after 20 years of Progress
Albert Gjedde, Dean F. Wong, Pedro Rosa-Neto, and Paul Cumming

Mitochondrial Dysfunction in Bipolar Disorder: From ^{31}P-Magnetic Resonance Spectroscopic Findings to Their Molecular Mechanisms
Tadafumi Kato

Large-Scale Microarray Studies of Gene Expression in Multiple Regions of the Brain in Schizophrenia and Alzeimer's Disease
Pavel L. Katsel, Kenneth L. Davis, and Vahram Haroutunian

Regulation of Serotonin 2C Receptor PRE-mRNA Editing By Serotonin
Claudia Schmauss

The Dopamine Hypothesis of Drug Addiction: Hypodopaminergic State
Miriam Melis, Saturnino Spiga, and Marco Diana

Human and Animal Spongiform Encephalopathies are Autoimmune Diseases: A Novel Theory and Its supporting Evidence
Bao Ting Zhu

Adenosine and Brain Function
Bertil B. Fredholm, Jiang-Fan Chen, Rodrigo A. Cunha, Per Svenningsson, and Jean-Marie Vaugeois

INDEX

Volume 64

Section I. The Cholinergic System
John Smythies

Section II. The Dopamine System
John Symythies

Section III. The Norepinephrine System
John Smythies

Section IV. The Adrenaline System
John Smythies

Section V. Serotonin System
John Smythies

INDEX

Volume 65

Insulin Resistance: Causes and Consequences
Zachary T. Bloomgarden

Antidepressant-Induced Manic Conversion: A Developmentally Informed Synthesis of the Literature
Christine J. Lim, James F. Leckman, Christopher Young, and AndrÉs Martin

Sites of Alcohol and Volatile Anesthetic Action on Glycine Receptors
Ingrid A. Lobo and R. Adron Harris

Role of the Orbitofrontal Cortex in Reinforcement Processing and Inhibitory Control: Evidence from Functional Magnetic Resonance Imaging Studies in Healthy Human Subjects
Rebecca Elliott and Bill Deakin

Common Substrates of Dysphoria in Stimulant Drug Abuse and Primary Depression: Therapeutic Targets
Kate Baicy, Carrie E. Bearden, John Monterosso, Arthur L. Brody, Andrew J. Isaacson, and Edythe D. London

The Role of cAMP Response Element–Binding Proteins in Mediating Stress-Induced Vulnerability to Drug Abuse
Arati Sadalge Kreibich and Julie A. Blendy

G-Protein–Coupled Receptor Deorphanizations
Yumiko Saito and Olivier Civelli

Mechanistic Connections Between Glucose/Lipid Disturbances and Weight Gain Induced by Antipsychotic Drugs
Donard S. Dwyer, Dallas Donohoe, Xiao-Hong Lu, and Eric J. Aamodt

Serotonin Firing Activity as a Marker for Mood Disorders: Lessons from Knockout Mice
Gabriella Gobbi

INDEX

Volume 66

Brain Atlases of Normal and Diseased Populations
Arthur W. Toga and Paul M. Thompson

Neuroimaging Databases as a Resource for Scientific Discovery
John Darrell Van Horn, John Wolfe, Autumn Agnoli, Jeffrey Woodward, Michael Schmitt, James Dobson, Sarene Schumacher, and Bennet Vance

Modeling Brain Responses
Karl J. Friston, William Penny, and Olivier David

Voxel-Based Morphometric Analysis Using Shape Transformations
Christos Davatzikos

The Cutting Edge of fMRI and High-Field fMRI
Dae-Shik Kim

Quantification of White Matter Using Diffusion-Tensor Imaging
Hae-Jeong Park

Perfusion fMRI for Functional Neuroimaging
Geoffrey K. Aguirre, John A. Detre, and Jiongjiong Wang

Functional Near-Infrared Spectroscopy: Potential and Limitations in Neuroimaging Studies
Yoko Hoshi

Neural Modeling and Functional Brain Imaging: The Interplay Between the Data-Fitting and Simulation Approaches
Barry Horwitz and Michael F. Glabus

Combined EEG and fMRI Studies of Human Brain Function
V. Menon and S. Crottaz-Herbette

INDEX

Volume 67

Distinguishing Neural Substrates of Heterogeneity Among Anxiety Disorders
Jack B. Nitschke and Wendy Heller

Neuroimaging in Dementia
K. P. Ebmeier, C. Donaghey, and N. J. Dougall

Prefrontal and Anterior Cingulate Contributions to Volition in Depression
Jack B. Nitschke and Kristen L. Mackiewicz

Functional Imaging Research in Schizophrenia
H. Tost, G. Ende, M. Ruf, F. A. Henn, and A. Meyer-Lindenberg

Neuroimaging in Functional Somatic Syndromes
Patrick B. Wood

Neuroimaging in Multiple Sclerosis
Alireza Minagar, Eduardo Gonzalez-Toledo, James Pinkston, and Stephen L. Jaffe

Stroke
Roger E. Kelley and Eduardo Gonzalez-Toledo

Functional MRI in Pediatric Neurobehavioral Disorders
Michael Seyffert and F. Xavier Castellanos

Structural MRI and Brain Development
Paul M. Thompson, Elizabeth R. Sowell, Nitin Gogtay, Jay N. Giedd, Christine N. Vidal, Kiralee M. Hayashi, Alex Leow, Rob Nicolson, Judith L. Rapoport, and Arthur W. Toga

Neuroimaging and Human Genetics
Georg Winterer, Ahmad R. Hariri, David Goldman, and Daniel R. Weinberger

Neuroreceptor Imaging in Psychiatry: Theory and Applications
W. Gordon Frankle, Mark Slifstein, Peter S. Talbot, and Marc Laruelle

INDEX

Volume 68

Fetal Magnetoencephalography: Viewing the Developing Brain *In Utero*
Hubert Preissl, Curtis L. Lowery, and Hari Eswaran

Magnetoencephalography in Studies of Infants and Children
Minna Huotilainen

Let's Talk Together: Memory Traces Revealed by Cooperative Activation in the Cerebral Cortex
Jochen Kaiser, Susanne Leiberg, and Werner Lutzenberger

Human Communication Investigated With Magnetoencephalography: Speech, Music, and Gestures
Thomas R. Knösche, Burkhard Maess, Akinori Nakamura, and Angela D. Friederici

Combining Magnetoencephalography and Functional Magnetic Resonance Imaging
Klaus Mathiak and Andreas J. Fallgatter

Beamformer Analysis of MEG Data
Arjan Hillebrand and Gareth R. Barnes

Functional Connectivity Analysis in Magnetoencephalography
Alfons Schnitzler and Joachim Gross

Human Visual Processing as Revealed by Magnetoencephalographys
Yoshiki Kaneoke, Shoko Watanabe, and Ryusuke Kakigi

A Review of Clinical Applications of Magnetoencephalography
Andrew C. Papanicolaou, Eduardo M. Castillo, Rebecca Billingsley-Marshall, Ekaterina Pataraia, and Panagiotis G. Simos

INDEX

Volume 69

Nematode Neurons: Anatomy and Anatomical Methods in Caenorhabditis elegans
David H. Hall, Robyn Lints, and Zeynep Altun

Investigations of Learning and Memory in Caenorhabditis elegans
Andrew C. Giles, Jacqueline K. Rose, and Catharine H. Rankin

Neural Specification and Differentiation
Eric Aamodt and Stephanie Aamodt

Sexual Behavior of the Caenorhabditis elegans Male
Scott W. Emmons

The Motor Circuit
Stephen E. Von Stetina, Millet Treinin, and David M. Miller III

Mechanosensation in Caenorhabditis elegans
Robert O'Hagan and Martin Chalfie

Volume 70

Spectral Processing by the Peripheral Auditory System Facts and Models
Enrique A. Lopez-Poveda

Basic Psychophysics of Human Spectral Processing
Brian C. J. Moore

Across-Channel Spectral Processing
John H. Grose, Joseph W. Hall III, and Emily Buss

Speech and Music Have Different Requirements for Spectral Resolution
Robert V. Shannon

Non-Linearities and the Representation of Auditory Spectra
Eric D. Young, Jane J. Yu, and Lina A. J. Reiss

Spectral Processing in the Inferior Colliculus
Kevin A. Davis

Neural Mechanisms for Spectral Analysis in the Auditory Midbrain, Thalamus, and Cortex
Monty A. Escabí and Heather L. Read

Spectral Processing in the Auditory Cortex
Mitchell L. Sutter

Processing of Dynamic Spectral Properties of Sounds
Adrian Rees and Manuel S. Malmierca

Representations of Spectral Coding in the Human Brain
Deborah A. Hall, PhD

Spectral Processing and Sound Source Determination
Donal G. Sinex

Spectral Information in Sound Localization
Simon Carlile, Russell Martin, and Ken McAnally

Plasticity of Spectral Processing
Dexter R. F. Irvine and Beverly A. Wright

Spectral Processing In Cochlear Implants
Colette M. McKay

INDEX

Volume 71

Autism: Neuropathology, Alterations of the GABAergic System, and Animal Models
Christoph Schmitz, Imke A. J. van Kooten, Patrick R. Hof, Herman van Engeland, Paul H. Patterson, and Harry W. M. Steinbusch

The Role of GABA in the Early Neuronal Development
Marta Jelitai and Emı'lia Madarasz

GABAergic Signaling in the Developing Cerebellum
Chitoshi Takayama

Insights into GABA Functions in the Developing Cerebellum
Mo'nica L. Fiszman

Role of GABA in the Mechanism of the Onset of Puberty in Non-Human Primates
Ei Terasawa

Rett Syndrome: A Rosetta Stone for Understanding the Molecular Pathogenesis of Autism
Janine M. LaSalle, Amber Hogart, and Karen N. Thatcher

GABAergic Cerebellar System in Autism: A Neuropathological and Developmental Perspective
Gene J. Blatt

Reelin Glycoprotein in Autism and Schizophrenia
S. Hossein Fatemi

Is There A Connection Between Autism, Prader-Willi Syndrome, Catatonia, and GABA?
Dirk M. Dhossche, Yaru Song, and Yiming Liu

Alcohol, GABA Receptors, and Neurodevelopmental Disorders
Ujjwal K. Rout

Effects of Secretin on Extracellular GABA and Other Amino Acid Concentrations in the Rat Hippocampus
Hans-Willi Clement, Alexander Pschibul, and Eberhard Schulz

Predicted Role of Secretin and Oxytocin in the Treatment of Behavioral and Developmental Disorders: Implications for Autism
Martha G. Welch and David A. Ruggiero

Immunological Findings in Autism
Hari Har Parshad Cohly and Asit Panja

Correlates of Psychomotor Symptoms in Autism
Laura Stoppelbein, Sara Sytsma-Jordan, and Leilani Greening

GABRB3 Gene Deficient Mice: A Potential Model of Autism Spectrum Disorder
Timothy M. DeLorey

The Reeler Mouse: Anatomy of a Mutant
Gabriella D'Arcangelo

Shared Chromosomal Susceptibility Regions Between Autism and Other Mental Disorders
Yvon C. Chagnon index

Index

Volume 72

Classification Matters for Catatonia and Autism in Children
Klaus-Jürgen Neumärker

A Systematic Examination of Catatonia-Like Clinical Pictures in Autism Spectrum Disorders
Lorna Wing and Amitta Shah

Catatonia in Individuals with Autism Spectrum Disorders in Adolescence and Early Adulthood: A Long-Term Prospective Study
Masataka Ohta, Yukiko Kano, and Yoko Nagai

Are Autistic and Catatonic Regression Related? A Few Working Hypotheses Involving GABA, Purkinje Cell Survival, Neurogenesis, and ECT
Dirk Marcel Dhossche and Ujjwal Rout

Psychomotor Development and Psychopathology in Childhood
Dirk M. J. De Raeymaecker

The Importance of Catatonia and Stereotypies in Autistic Spectrum Disorders
Laura Stoppelbein, Leilani Greening, and Angelina Kakooza

Prader–Willi Syndrome: Atypical Psychoses and Motor Dysfunctions
Willem M. A. Verhoeven and Siegfried Tuinier

Towards a Valid Nosography and Psychopathology of Catatonia in Children and Adolescents
David Cohen

Is There a Common Neuronal Basis for Autism and Catatonia?
Dirk Marcel Dhossche, Brendan T. Carroll, and Tressa D. Carroll

Shared Susceptibility Region on Chromosome 15 Between Autism and Catatonia
Yvon C. Chagnon

Current Trends in Behavioral Interventions for Children with Autism
Dorothy Scattone and Kimberly R. Knight

Case Reports with a Child Psychiatric Exploration of Catatonia, Autism, and Delirium
Jan N. M. Schieveld

ECT and the Youth: Catatonia in Context
Frank K. M. Zaw

Catatonia in Autistic Spectrum Disorders: A Medical Treatment Algorithm
Max Fink, Michael A. Taylor, and Neera Ghaziuddin

Psychological Approaches to Chronic Catatonia-Like Deterioration in Autism Spectrum Disorders
Amitta Shah and Lorna Wing

Section V: Blueprints

Blueprints for the Assessment, Treatment, and Future Study of Catatonia in Autism Spectrum Disorders
Dirk Marcel, Dhossche, Amitta Shah, and Lorna Wing

Index

Volume 73

Chromosome 22 Deletion Syndrome and Schizophrenia
Nigel M. Williams, Michael C. O'Donovan, and Michael J. Owen

Characterization of Proteome of Human Cerebrospinal Fluid
Jing Xu, Jinzhi Chen, Elaine R. Peskind, Jinghua Jin, Jimmy Eng, Catherine Pan, Thomas J. Montine, David R. Goodlett, and Jing Zhang

Hormonal Pathways Regulating Intermale and Interfemale Aggression
Neal G. Simon, Qianxing Mo, Shan Hu, Carrie Garippa, and Shi-Fang Lu

Neuronal GAP Junctions: Expression, Function, and Implications for Behavior
Clinton B. McCracken and David C. S. Roberts

Effects of Genes and Stress on the Neurobiology of Depression
J. John Mann and Dianne Currier

Quantitative Imaging with the Micropet Small-Animal Pet Tomograph
Paul Vaska, Daniel J. Rubins, David L. Alexoff, and Wynne K. Schiffer

Understanding Myelination through Studying its Evolution
Rüdiger Schweigreiter, Betty I. Roots, Christine Bandtlow, and Robert M. Gould

INDEX

Volume 74

Evolutionary Neurobiology and Art
C. U. M. Smith

Section I: Visual Aspects

Perceptual Portraits
Nicholas Wade

The Neuropsychology of Visual Art: Conferring Capacity
Anjan Chatterjee

Vision, Illusions, and Reality
Christopher Kennard

Localization in the Visual Brain
George K. York

Section II: Episodic Disorders

Neurology, Synaesthesia, and Painting
Amy Ione

Fainting in Classical Art
Philip Smith

Migraine Art in the Internet: A Study of 450 Contemporary Artists
Klaus Podoll

Sarah Raphael's Migraine with Aura as Inspiration for the Foray of Her Work into Abstraction
Klaus Podoll and Debbie Ayles

The Visual Art of Contemporary Artists with Epilepsy
Steven C. Schachter

Section III: Brain Damage

Creativity in Painting and Style in Brain-Damaged Artists
Julien Bogousslavsky

Artistic Changes in Alzheimer's Disease
Sebastian J. Crutch and Martin N. Rossor

Section IV: Cerebrovascular Disease

Stroke in Painters
H. Bäzner and M. Hennerici

Visuospatial Neglect in Lovis Corinth's Self-Portraits
Olaf Blanke

Art, Constructional Apraxia, and the Brain
Louis Caplan

Section V: Genetic Diseases

Neurogenetics in Art
Alan E. H. Emery

A Naïve Artist of St Ives
F. Clifford Rose

Van Gogh's Madness
F. Clifford Rose

Absinthe, The Nervous System and Painting
Tiina Rekand

Section VI: Neurologists as Artists

Sir Charles Bell, KGH, FRS, FRSE (1774–1842)
Christopher Gardner-Thorpe

Section VII: Miscellaneous

Peg Leg Frieda
Espen Dietrichs

The Deafness of Goya (1746–1828)
F. Clifford Rose

INDEX

Volume 75

Introduction on the Use of the *Drosophila* Embryonic/Larval Neuromuscular Junction as a Model System to Study Synapse Development and Function, and a Brief Summary of Pathfinding and Target Recognition
Catalina Ruiz-Cañada and Vivian Budnik

Development and Structure of Motoneurons
Matthias Landgraf and Stefan Thor

The Development of the *Drosophila* Larval Body Wall Muscles
Karen Beckett and Mary K. Baylies

Organization of the Efferent System and Structure of Neuromuscular Junctions in *Drosophila*
Andreas Prokop

Development of Motoneuron Electrical Properties and Motor Output
Richard A. Baines

Transmitter Release at the Neuromuscular Junction
Thomas L. Schwarz

Vesicle Trafficking and Recycling at the Neuromuscular Junction: Two Pathways for Endocytosis
Yoshiaki Kidokoro

Glutamate Receptors at the *Drosophila* Neuromuscular Junction
Aaron DiAntonio

Scaffolding Proteins at the *Drosophila* Neuromuscular Junction
Bulent Ataman, Vivian Budnik, and Ulrich Thomas

Synaptic Cytoskeleton at the Neuromuscular Junction
Catalina Ruiz-Cañada and Vivian Budnik

Plasticity and Second Messengers During Synapse Development
Leslie C. Griffith and Vivian Budnik

Retrograde Signaling that Regulates Synaptic Development and Function at the *Drosophila* Neuromuscular Junction
Guillermo Marqués and Bing Zhang

Activity-Dependent Regulation of Transcription During Development of Synapses
Subhabrata Sanyal and Mani Ramaswami

Experience-Dependent Potentiation of Larval Neuromuscular Synapses
Christoph M. Schuster

Selected Methods for the Anatomical Study of *Drosophila* Embryonic and Larval Neuromuscular Junctions
Vivian Budnik, Michael Gorczyca, and Andreas Prokop

INDEX

Volume 76

Section I: Physiological Correlates of Freud's Theories

The ID, the Ego, and the Temporal Lobe
Shirley M. Ferguson and Mark Rayport

ID, Ego, and Temporal Lobe Revisited
Shirley M. Ferguson and Mark Rayport

Section II: Stereotaxic Studies

Olfactory Gustatory Responses Evoked by Electrical Stimulation of Amygdalar Region in Man Are Qualitatively Modifiable by Interview Content: Case Report and Review
Mark Rayport, Sepehr Sani, and Shirley M. Ferguson

Section III: Controversy in Definition of Behavioral Disturbance

Pathogenesis of Psychosis in Epilepsy. The "Seesaw" Theory: Myth or Reality?
Shirley M. Ferguson and Mark Rayport

Section IV: Outcome of Temporal Lobectomy

Memory Function After Temporal Lobectomy for Seizure Control: A Comparative Neuropsy chiatric and Neuropsychological Study
Shirley M. Ferguson, A. John McSweeny, and Mark Rayport

Life After Surgery for Temporolimbic Seizures
Shirley M. Ferguson, Mark Rayport, and Carolyn A. Schell

Appendix I
Mark Rayport

Appendix II: Conceptual Foundations of Studies of Patients Undergoing Temporal Lobe Surgery for Seizure Control
Mark Rayport

INDEX

Volume 77

Regenerating the Brain
David A. Greenberg and Kunlin Jin

Serotonin and Brain: Evolution, Neuroplasticity, and Homeostasis
Efrain C. Azmitia

Therapeutic Approaches to Promoting Axonal Regeneration in the Adult Mammalian Spinal Cord
Sari S. Hannila, Mustafa M. Siddiq, and Marie T. Filbin

Evidence for Neuroprotective Effects of Antipsychotic Drugs: Implications for the Pathophysiology and Treatment of Schizophrenia
Xin-Min Li and Haiyun Xu

Neurogenesis and Neuroenhancement in the Pathophysiology and Treatment of Bipolar Disorder
Robert J. Schloesser, Guang Chen, and Husseini K. Manji

Neuroreplacement, Growth Factor, and Small Molecule Neurotrophic Approaches for Treating Parkinson's Disease
Michael J. O'Neill, Marcus J. Messenger, Viktor Lakics, Tracey K. Murray, Eric H. Karran, Philip G. Szekeres, Eric S. Nisenbaum, and Kalpana M. Merchant

Using *Caenorhabditis elegans* Models of Neurodegenerative Disease to Identify Neuroprotective Strategies
Brian Kraemer and Gerard D. Schellenberg

Neuroprotection and Enhancement of Neurite Outgrowth With Small Molecular Weight Compounds From Screens of Chemical Libraries
Donard S. Dwyer and Addie Dickson

INDEX

Volume 78

Neurobiology of Dopamine in Schizophrenia
Olivier Guillin, Anissa Abi-Dargham, and Marc Laruelle

The Dopamine System and the Pathophysiology of Schizophrenia: A Basic Science Perspective
Yukiori Goto and Anthony A. Grace

Glutamate and Schizophrenia: Phencyclidine, *N*-methyl-D-aspartate Receptors, and Dopamine–Glutamate Interactions
Daniel C. Javitt

Deciphering the Disease Process of Schizophrenia: The Contribution of Cortical GABA Neurons
David A. Lewis and Takanori Hashimoto

Alterations of Serotonin Transmission in Schizophrenia
Anissa Abi-Dargham

Serotonin and Dopamine Interactions in Rodents and Primates: Implications for Psychosis and Antipsychotic Drug Development
Gerard J. Marek

Cholinergic Circuits and Signaling in the Pathophysiology of Schizophrenia
Joshua A. Berman, David A. Talmage, and Lorna W. Role

Schizophrenia and the α7 Nicotinic Acetylcholine Receptor
Laura F. Martin and Robert Freedman

Histamine and Schizophrenia
Jean-Michel Arrang

Cannabinoids and Psychosis
Deepak Cyril D'Souza

Involvement of Neuropeptide Systems in Schizophrenia: Human Studies
Ricardo Cáceda, Becky Kinkead, and Charles B. Nemeroff

Brain-Derived Neurotrophic Factor in Schizophrenia and Its Relation with Dopamine
Olivier Guillin, Caroline Demily, and Florence Thibaut

Schizophrenia Susceptibility Genes: In Search of a Molecular Logic and Novel Drug Targets for a Devastating Disorder
Joseph A. Gogos

INDEX

Volume 79

The Destructive Alliance: Interactions of Leukocytes, Cerebral Endothelial Cells, and the Immune Cascade in Pathogenesis of Multiple Sclerosis
Alireza Minagar, April Carpenter, and J. Steven Alexander

Role of B Cells in Pathogenesis of Multiple Sclerosis
Behrouz Nikbin, Mandana Mohyeddin Bonab, Farideh Khosravi, and Fatemeh Talebian

The Role of CD4 T Cells in the Pathogenesis of Multiple Sclerosis
Tanuja Chitnis

The CD8 T Cell in Multiple Sclerosis: Suppressor Cell or Mediator of Neuropathology?
Aaron J. Johnson, Georgette L. Suidan, Jeremiah McDole, and Istvan Pirko

Immunopathogenesis of Multiple Sclerosis
Smriti M. Agrawal and V. Wee Yong

Molecular Mimicry in Multiple Sclerosis
Jane E. Libbey, Lori L. McCoy, and Robert S. Fujinami

Molecular "Negativity" May Underlie Multiple Sclerosis: Role of the Myelin Basic Protein Family in the Pathogenesis of MS
Abdiwahab A. Musse and George Harauz

Microchimerism and Stem Cell Transplantation in Multiple Sclerosis
Behrouz Nikbin, Mandana Mohyeddin Bonab, and Fatemeh Talebian

The Insulin-Like Growth Factor System in Multiple Sclerosis
Daniel Chesik, Nadine Wilczak, and Jacques De Keyser

Cell-Derived Microparticles and Exosomes in Neuroinflammatory Disorders
Lawrence L. Horstman, Wenche Jy, Alireza Minagar, Carlos J. Bidot, Joaquin J. Jimenez, J. Steven Alexander, and Yeon S. Ahn

Multiple Sclerosis in Children: Clinical, Diagnostic, and Therapeutic Aspects
Kevin Rostásy

Migraine in Multiple Sclerosis
Debra G. Elliott

Multiple Sclerosis as a Painful Disease
Meghan Kenner, Uma Menon, and Debra Elliott

Multiple Sclerosis and Behavior
James B. Pinkston, Anita Kablinger, and Nadejda Alekseeva

Cerebrospinal Fluid Analysis in Multiple Sclerosis
Francisco A. Luque and Stephen L. Jaffe

Multiple Sclerosis in Isfahan, Iran
Mohammad Saadatnia, Masoud Etemadifar, and Amir Hadi Maghzi

Gender Issues in Multiple Sclerosis
Robert N. Schwendimann and Nadejda Alekseeva

Differential Diagnosis of Multiple Sclerosis
Halim Fadil, Roger E. Kelley, and Eduardo Gonzalez-Toledo

Prognostic Factors in Multiple Sclerosis
Roberto Bergamaschi

Neuroimaging in Multiple Sclerosis
Robert Zivadinov and Jennifer L. Cox

Detection of Cortical Lesions Is Dependent on Choice of Slice Thickness in Patients with Multiple Sclerosis
Ondrej Dolezal, Michael G. Dwyer, Dana Horakova, Eva Havrdova, Alireza Minagar, Srivats Balachandran, Niels Bergsland, Zdenek Seidl, Manuela Vaneckova, David Fritz, Jan Krasensky, and Robert Zivadinov

The Role of Quantitative Neuroimaging Indices in the Differentiation of Ischemia from Demyelination: An Analytical Study with Case Presentation
Romy Hoque, Christina Ledbetter, Eduardo Gonzalez-Toledo, Vivek Misra, Uma Menon, Meghan Kenner, Alejandro A. Rabinstein, Roger E. Kelley, Robert Zivadinov, and Alireza Minagar

HLA-DRB1*1501, -DQB1*0301, -DQB1*0302, -DQB1*0602, and -DQB1*0603 Alleles Are Associated with More Severe Disease Outcome on MRI in Patients with Multiple Sclerosis
Robert Zivadinov, Laura Uxa, Alessio Bratina, Antonio Bosco, Bhooma Srinivasaraghavan, Alireza Minagar, Maja Ukmar, Su yen Benedetto, and Marino Zorzon

Glatiramer Acetate: Mechanisms of Action in Multiple Sclerosis
Tjalf Ziemssen and Wiebke Schrempf

Evolving Therapies for Multiple Sclerosis
Elena Korniychuk, John M. Dempster, Eileen O'Connor, J. Steven Alexander, Roger E. Kelley, Meghan Kenner, Uma Menon, Vivek Misra, Romy Hoque, Eduardo C. Gonzalez-Toledo, Robert N. Schwendimann, Stacy Smith, and Alireza Minagar

Remyelination in Multiple Sclerosis
Divya M. Chari

Trigeminal Neuralgia: A Modern-Day Review
Kelly Hunt and Ravish Patwardhan

Optic Neuritis and the Neuro-Ophthalmology of Multiple Sclerosis
Paramjit Kaur and Jeffrey L. Bennett

Neuromyelitis Optica: New Findings on Pathogenesis
Dean M. Wingerchuk

INDEX

Volume 80

Epilepsy in the Elderly: Scope of the Problem
Ilo E. Leppik

Animal Models in Gerontology Research
Nancy L. Nadon

Animal Models of Geriatric Epilepsy
Lauren J. Murphree, Lynn M. Rundhaugen, and Kevin M. Kelly

Life and Death of Neurons in the Aging Cerebral Cortex
John H. Morrison and Patrick R. Hof

An In Vitro Model of Stroke-Induced Epilepsy: Elucidation of the Roles of Glutamate and Calcium in the Induction and Maintenance of Stroke-Induced Epileptogenesis
Robert J. DeLorenzo, David A. Sun, Robert E. Blair, and Sompong Sambati

Mechanisms of Action of Antiepileptic Drugs
H. Steve White, Misty D. Smith, and Karen S. Wilcox

Epidemiology and Outcomes of Status Epilepticus in the Elderly
Alan R. Towne

Diagnosing Epilepsy in the Elderly
R. Eugene Ramsay, Flavia M. Macias, and A. James Rowan

Pharmacoepidemiology in Community-Dwelling Elderly Taking Antiepileptic Drugs
Dan R. Berlowitz and Mary Jo V. Pugh

Use of Antiepileptic Medications in Nursing Homes
Judith Garrard, Susan L. Harms, Lynn E. Eberly, and Ilo E. Leppik

Differential Diagnosis of Multiple Sclerosis
Halim Fadil, Roger E. Kelley, and Eduardo Gonzalez-Toledo

Prognostic Factors in Multiple Sclerosis
Roberto Bergamaschi

Neuroimaging in Multiple Sclerosis
Robert Zivadinov and Jennifer L. Cox

Detection of Cortical Lesions Is Dependent on Choice of Slice Thickness in Patients with Multiple Sclerosis
Ondrej Dolezal, Michael G. Dwyer, Dana Horakova, Eva Havrdova, Alireza Minagar, Srivats Balachandran, Niels Bergsland, Zdenek Seidl, Manuela Vaneckova, David Fritz, Jan Krasensky, and Robert Zivadinov

The Role of Quantitative Neuroimaging Indices in the Differentiation of Ischemia from Demyelination: An Analytical Study with Case Presentation
Romy Hoque, Christina Ledbetter, Eduardo Gonzalez-Toledo, Vivek Misra, Uma Menon, Meghan Kenner, Alejandro A. Rabinstein, Roger E. Kelley, Robert Zivadinov, and Alireza Minagar

HLA-DRB1*1501, -DQB1*0301,-DQB1 *0302,-DQB1*0602, and -DQB1*0603 Alleles Are Associated with More Severe Disease Outcome on MRI in Patients with Multiple Sclerosis
Robert Zivadinov, Laura Uxa, Alessio Bratina, Antonio Bosco, Bhooma Srinivasaraghavan, Alireza Minagar, Maja Ukmar, Su yen Benedetto, and Marino Zorzon

Glatiramer Acetate: Mechanisms of Action in Multiple Sclerosis
Tjalf Ziemssen and Wiebke Schrempf

Evolving Therapies for Multiple Sclerosis
Elena Korniychuk, John M. Dempster, Eileen O'Connor, J. Steven Alexander, Roger E. Kelley, Meghan Kenner, Uma Menon, Vivek Misra, Romy Hoque, Eduardo C. Gonzalez-Toledo, Robert N. Schwendimann, Stacy Smith, and Alireza Minagar

Remyelination in Multiple Sclerosis
Divya M. Chari

Trigeminal Neuralgia: A Modern-Day Review
Kelly Hunt and Ravish Patwardhan

Optic Neuritis and the Neuro-Ophthalmology of Multiple Sclerosis
Paramjit Kaur and Jeffrey L. Bennett

Neuromyelitis Optica: New Findings on Pathogenesis
Dean M. Wingerchuk

INDEX

Volume 81

Epilepsy in the Elderly: Scope of the Problem
Ilo E. Leppik

Animal Models in Gerontology Research
Nancy L. Nadon

Animal Models of Geriatric Epilepsy
Lauren J. Murphree, Lynn M. Rundhaugen, and Kevin M. Kelly

Life and Death of Neurons in the Aging Cerebral Cortex
John H. Morrison and Patrick R. Hof

An In Vitro Model of Stroke-Induced Epilepsy: Elucidation of the Roles of Glutamate and Calcium in the Induction and Maintenance of Stroke-Induced Epileptogenesis
Robert J. DeLorenzo, David A. Sun, Robert E. Blair, and Sompong Sambati

Mechanisms of Action of Antiepileptic Drugs
H. Steve White, Misty D. Smith, and Karen S. Wilcox

Epidemiology and Outcomes of Status Epilepticus in the Elderly
Alan R. Towne

Diagnosing Epilepsy in the Elderly
R. Eugene Ramsay, Flavia M. Macias, and A. James Rowan

Pharmacoepidemiology in Community-Dwelling Elderly Taking Antiepileptic Drugs
Dan R. Berlowitz and Mary Jo V. Pugh

Use of Antiepileptic Medications in Nursing Homes
Judith Garrard, Susan L. Harms, Lynn E. Eberly, and Ilo E. Leppik

Age-Related Changes in Pharmacokinetics: Predictability and Assessment Methods
Emilio Perucca

Factors Affecting Antiepileptic Drug Pharmacokinetics in Community-Dwelling Elderly
James C. Cloyd, Susan Marino, and Angela K. Birnbaum

Pharmacokinetics of Antiepileptic Drugs in Elderly Nursing Home Residents
Angela K. Birnbaum

The Impact of Epilepsy on Older Veterans
Mary Jo V. Pugh, Dan R. Berlowitz, and Lewis Kazis

Risk and Predictability of Drug Interactions in the Elderly
René H. Levy and Carol Collins

Outcomes in Elderly Patients With Newly Diagnosed and Treated Epilepsy
Martin J. Brodie and Linda J. Stephen

Recruitment and Retention in Clinical Trials of the Elderly
Flavia M. Macias, R. Eugene Ramsay, and A. James Rowan

Treatment of Convulsive Status Epilepticus
David M. Treiman

Treatment of Nonconvulsive Status Epilepticus
Matthew C. Walker

Antiepileptic Drug Formulation and Treatment in the Elderly: Biopharmaceutical Considerations
Barry E. Gidal

INDEX

Volume 82

Inflammatory Mediators Leading to Protein Misfolding and Uncompetitive/Fast Off-Rate Drug Therapy for Neurodegenerative Disorders
Stuart A. Lipton, Zezong Gu, and Tomohiro Nakamura

Innate Immunity and Protective Neuroinflammation: New Emphasis on the Role of Neuroimmune Regulatory Proteins
M. Griffiths, J. W. Neal, and P. Gasque

Glutamate Release from Astrocytes in Physiological Conditions and in Neurodegenerative Disorders Characterized by Neuroinflammation
Sabino Vesce, Daniela Rossi, Liliana Brambilla, and Andrea Volterra

The High-Mobility Group Box 1 Cytokine Induces Transporter-Mediated Release of Glutamate from Glial Subcellular Particles (Gliosomes) Prepared from *In Situ*-Matured Astrocytes
Giambattista Bonanno, Luca Raiteri, Marco Milanese, Simona Zappettini, Edon Melloni, Marco Pedrazzi, Mario Passalacqua, Carlo Tacchetti, Cesare Usai, and Bianca Sparatore

The Role of Astrocytes and Complement System in Neural Plasticity
Milos Pekny, Ulrika Wilhelmsson, Yalda Rahpeymai Bogestål, and Marcela Pekna

New Insights into the Roles of Metalloproteinases in Neurodegeneration and Neuroprotection
A. J. Turner and N. N. Nalivaeva

Relevance of High-Mobility Group Protein Box 1 to Neurodegeneration
Silvia Fossati and Alberto Chiarugi

Early Upregulation of Matrix Metalloproteinases Following Reperfusion Triggers Neuro-inflammatory Mediators in Brain Ischemia in Rat
Diana Amantea, Rossella Russo, Micaela Gliozzi, Vincenza Fratto, Laura Berliocchi, G. Bagetta, G. Bernardi, and M. Tiziana Corasaniti

The (Endo)Cannabinoid System in Multiple Sclerosis and Amyotrophic Lateral Sclerosis
Diego Centonze, Silvia Rossi, Alessandro Finazzi-Agrò, Giorgio Bernardi, and Mauro Maccarrone

Chemokines and Chemokine Receptors: Multi-purpose Players in Neuroinflammation
Richard M. Ransohoff, LiPing Liu, and Astrid E. Cardona

Systemic and Acquired Immune Responses in Alzheimer's Disease
Markus Britschgi and Tony Wyss-Coray

Neuroinflammation in Alzheimer's Disease and Parkinson's Disease: Are Microglia Pathogenic in Either Disorder?
Joseph Rogers, Diego Mastroeni, Brian Leonard, Jeffrey Joyce, and Andrew Grover

Cytokines and Neuronal Ion Channels in Health and Disease
Barbara Viviani, Fabrizio Gardoni, and Marina Marinovich

Cyclooxygenase-2, Prostaglandin E_2, and Microglial Activation in Prion Diseases
Luisa Minghetti and Maurizio Pocchiari

Glia Proinflammatory Cytokine Upregulation as a Therapeutic Target for Neurodegenerative Diseases: Function-Based and Target-Based Discovery Approaches
Linda J. Van Eldik, Wendy L. Thompson, Hantamalala Ralay Ranaivo, Heather A. Behanna, and D. Martin Watterson

Oxidative Stress and the Pathogenesis of Neuro-degenerative Disorders
Ashley Reynolds, Chad Laurie, R. Lee Mosley, and Howard E. Gendelman

Differential Modulation of Type 1 and Type 2 Cannabinoid Receptors Along the Neuroimmune Axis
Sergio Oddi, Paola Spagnuolo, Monica Bari, Antonella D'Agostino, and Mauro Maccarrone

Effects of the HIV-1 Viral Protein Tat on Central Neurotransmission: Role of Group I Meta-botropic Glutamate Receptors
Elisa Neri, Veronica Musante, and Anna Pittaluga

Evidence to Implicate Early Modulation of Inter-leukin-1β Expression in the Neuroprotection Afforded by 17β-Estradiol in Male Rats Undergone Transient Middle Cerebral Artery Occlusion
Olga Chiappetta, Micaela Gliozzi, Elisa Siviglia, Diana Amantea, Luigi A. Morrone, Laura Berliocchi, G. Bagetta, and M. Tiziana Corasaniti

A Role for Brain Cyclooxygenase-2 and Prosta-glandin-E_2 in Migraine: Effects of Nitroglycerin
Cristina Tassorelli, Rosaria Greco, Marie Therèse Armentero, Fabio Blandini, Giorgio Sandrini, and Giuseppe Nappi

The Blockade of K+-ATP Channels has Neuro-protective Effects in an *In Vitro* Model of Brain Ischemia
Robert Nisticò, Silvia Piccirilli, L. Sebastianelli, Giuseppe Nisticò, G. Bernardi, and N. B. Mercuri

Retinal Damage Caused by High Intraocular Pressure-Induced Transient Ischemia is Prevented by Coenzyme Q10 in Rat
Carlo Nucci, Rosanna Tartaglione, Angelica Cerulli, R. Mancino, A. Spanò, Federica Cavaliere, Laura Rombolà, G. Bagetta, M. Tiziana Corasaniti, and Luigi A. Morrone

Evidence Implicating Matrix Metalloproteinases in the Mechanism Underlying Accumulation of IL-1β and Neuronal Apoptosis in the Neocortex of HIV/gp120-Exposed Rats
Rossella Russo, Elisa Siviglia, Micaela Gliozzi, Diana Amantea, Annamaria Paoletti, Laura Berliocchi, G. Bagetta, and M. Tiziana Corasaniti

Neuroprotective Effect of Nitroglycerin in a Rodent Model of Ischemic Stroke: Evaluation of Bcl-2 Expression
Rosaria Greco, Diana Amantea, Fabio Blandini, Giuseppe Nappi, Giacinto Bagetta, M. Tiziana Corasaniti, and Cristina Tassorelli

INDEX

Volume 83

Gender Differences in Pharmacological Response
Gail D. Anderson

Epidemiology and Classification of Epilepsy: Gender Comparisons
John C. McHugh and Norman Delanty

Hormonal Influences on Seizures: Basic Neurobiology
Cheryl A. Frye

Catamenial Epilepsy
Patricia E. Penovich and Sandra Helmers

Epilepsy in Women: Special Considerations for Adolescents
Mary L. Zupanc and Sheryl Haut

Contraception in Women with Epilepsy: Pharmacokinetic Interactions, Contraceptive Options, and Management
Caryn Dutton and Nancy Foldvary-Schaefer

Reproductive Dysfunction in Women with Epilepsy: Menstrual Cycle Abnormalities, Fertility, and Polycystic Ovary Syndrome
Jürgen Bauer and Déirdre Cooper-Mahkorn

Sexual Dysfunction in Women with Epilepsy: Role of Antiepileptic Drugs and Psychotropic Medications
Mary A. Gutierrez, Romila Mushtaq, and Glen Stimmel

Pregnancy in Epilepsy: Issues of Concern
John DeToledo

Teratogenicity and Antiepileptic Drugs: Potential Mechanisms
Mark S. Yerby

Antiepileptic Drug Teratogenesis: What are the Risks for Congenital Malformations and Adverse Cognitive Outcomes?
Cynthia L. Harden

Teratogenicity of Antiepileptic Drugs: Role of Pharmacogenomics
Raman Sankar and Jason T. Lerner

Antiepileptic Drug Therapy in Pregnancy I: Gestation-Induced Effects on AED Pharmacokinetics
Page B. Pennell and Collin A. Hovinga

Antiepileptic Drug Therapy in Pregnancy II: Fetal and Neonatal Exposure
Collin A. Hovinga and Page B. Pennell

Seizures in Pregnancy: Diagnosis and Management
Robert L. Beach and Peter W. Kaplan

Management of Epilepsy and Pregnancy: An Obstetrical Perspective
Julian N. Robinson and Jane Cleary-Goldman

Pregnancy Registries: Strengths, Weaknesses, and Bias Interpretation of Pregnancy Registry Data
Marianne Cunnington and John Messenheimer

Bone Health in Women With Epilepsy: Clinical Features and Potential Mechanisms
Alison M. Pack and Thaddeus S. Walczak

Metabolic Effects of AEDs: Impact on Body Weight, Lipids and Glucose Metabolism
Raj D. Sheth and Georgia Montouris

Psychiatric Comorbidities in Epilepsy
W. Curt Lafrance, Jr., Andres M. Kanner, and Bruce Hermann

Issues for Mature Women with Epilepsy
Cynthia L. Harden

Pharmacodynamic and Pharmacokinetic Interactions of Psychotropic Drugs with Antiepileptic Drugs
Andres M. Kanner and Barry E. Gidal

Health Disparities in Epilepsy: How Patient-Oriented Outcomes in Women Differ from Men
Frank Gilliam

INDEX

Volume 84

Normal Brain Aging: Clinical, Immunological, Neuropsychological, and Neuroimaging Features
Maria T. Caserta, Yvonne Bannon, Francisco Fernandez, Brian Giunta, Mike R. Schoenberg, and Jun Tan

Subcortical Ischemic Cerebrovascular Dementia
Uma Menon and Roger E. Kelley

Cerebrovascular and Cardiovascular Pathology in Alzheimer's Disease
Jack C. de la Torre

Neuroimaging of Cognitive Impairments in Vascular Disease
Carol Di Perri, Turi O. Dalaker, Mona K. Beyer, and Robert Zivadinov

Contributions of Neuropsychology and Neuroimaging to Understanding Clinical Subtypes of Mild Cognitive Impairment
Amy J. Jak, Katherine J. Bangen, Christina E. Wierenga, Lisa Delano-Wood, Jody Corey-Bloom, and Mark W. Bondi

Proton Magnetic Resonance Spectroscopy in Dementias and Mild Cognitive Impairment
H. Randall Griffith, Christopher C. Stewart, and Jan A. den Hollander

Application of PET Imaging to Diagnosis of Alzheimer's Disease and Mild Cognitive Impairment
James M. Noble and Nikolaos Scarmeas

The Molecular and Cellular Pathogenesis of Dementia of the Alzheimer's Type: An Overview
Francisco A. Luque and Stephen L. Jaffe

Alzheimer's Disease Genetics: Current Status and Future Perspectives
Lars Bertram

Frontotemporal Lobar Degeneration: Insights from Neuropsychology and Neuroimaging
Andrea C. Bozoki and Muhammad U. Farooq

Lewy Body Dementia
Jennifer C. Hanson and Carol F. Lippa

Dementia in Parkinson's Disease
Bradley J. Robottom and William J. Weiner

Early Onset Dementia
Halim Fadil, Aimee Borazanci, Elhachmia Ait Ben Haddou, Mohamed Yahyaoui, Elena Korniychuk, Stephen L. Jaffe, and Alireza Minagar

Normal Pressure Hydrocephalus
Glen R. Finney

Reversible Dementias
Anahid Kabasakalian and Glen R. Finney

INDEX

Volume 85

Involvement of the Prefrontal Cortex in Problem Solving
Hajime Mushiake, Kazuhiro Sakamoto, Naohiro Saito, Toshiro Inui, Kazuyuki Aihara, and Jun Tanji

GluK1 Receptor Antagonists and Hippocampal Mossy Fiber Function
Robert Nisticò, Sheila Dargan, Stephen M. Fitzjohn, David Lodge, David E. Jane, Graham L. Collingridge, and Zuner A. Bortolotto

Monoamine Transporter as a Target Molecule for Psychostimulants
Ichiro Sora, BingJin Li, Setsu Fumushima, Asami Fukui, Yosefu Arime, Yoshiyuki Kasahara, Hiroaki Tomita, and Kazutaka Ikeda

Targeted Lipidomics as a Tool to Investigate Endocannabinoid Function
Giuseppe Astarita, Jennifer Geaga, Faizy Ahmed, and Daniele Piomelli

The Endocannabinoid System as a Target for Novel Anxiolytic and Antidepressant Drugs
Silvana Gaetani, Pasqua Dipasquale, Adele Romano, Laura Righetti, Tommaso Cassano, Daniele Piomelli, and Vincenzo Cuomo

$GABA_A$ Receptor Function and Gene Expression During Pregnancy and Postpartum
Giovanni Biggio, Maria Cristina Mostallino, Paolo Follesa, Alessandra Concas, and Enrico Sanna

Early Postnatal Stress and Neural Circuit Underlying Emotional Regulation
Machiko Matsumoto, Mitsuhiro Yoshioka, and Hiroko Togashi

Roles of the Histaminergic Neurotransmission on Methamphetamine-Induced Locomotor Sensitization and Reward: A Study of Receptors Gene Knockout Mice
Naoko Takino, Eiko Sakurai, Atsuo Kuramasu, Nobuyuki Okamura, and Kazuhiko Yanai

Developmental Exposure to Cannabinoids Causes Subtle and Enduring Neurofunctional Alterations
Patrizia Campolongo, Viviana Trezza, Maura Palmery, Luigia Trabace, and Vincenzo Cuomo

Neuronal Mechanisms for Pain-Induced Aversion: Behavioral Studies Using a Conditioned Place Aversion Test
Masabumi Minami

Bv8/Prokineticins and their Receptors: A New Pronociceptive System
Lucia Negri, Roberta Lattanzi, Elisa Giannini, Michela Canestrelli, Annalisa Nicotra, and Pietro Melchiorri

P2Y$_6$-Evoked Microglial Phagocytosis
Kazuhide Inoue, Schuichi Koizumi, Ayako Kataoka, Hidetoshi Tozaki-Saitoh, and Makoto Tsuda

PPAR and Pain
Takehiko Maeda and Shiroh Kishioka

Involvement of Inflammatory Mediators in Neuropathic Pain Caused by Vincristine
Norikazu Kiguchi, Takehiko Maeda, Yuka Kobayashi, Fumihiro Saika, and Shiroh Kishioka

Nociceptive Behavior Induced by the Endogenous Opioid Peptides Dynorphins in Uninjured Mice: Evidence with Intrathecal *N*-ethylmaleimide Inhibiting Dynorphin Degradation
Koichi Tan-No, Hiroaki Takahashi, Osamu Nakagawasai, Fukie Niijima, Shinobu Sakurada, Georgy Bakalkin, Lars Terenius, and Takeshi Tadano

Mechanism of Allodynia Evoked by Intrathecal Morphine-3-Glucuronide in Mice
Takaaki Komatsu, Shinobu Sakurada, Sou Katsuyama, Kengo Sanai, and Tsukasa Sakurada

(−)-Linalool Attenuates Allodynia in Neuropathic Pain Induced by Spinal Nerve Ligation in C57/Bl6 Mice
Laura Berliocchi, Rossella Russo, Alessandra Levato, Vincenza Fratto, Giacinto Bagetta, Shinobu Sakurada, Tsukasa Sakurada, Nicola Biagio Mercuri, and Maria Tiziana Corasaniti

Intraplantar Injection of Bergamot Essential Oil into the Mouse Hindpaw: Effects on Capsaicin-Induced Nociceptive Behaviors
Tsukasa Sakurada, Hikari Kuwahata, Soh Katsuyama, Takaaki Komatsu, Luigi A. Morrone, M. Tiziana Corasaniti, Giacinto Bagetta, and Shinobu Sakurada

New Therapy for Neuropathic Pain
Hirokazu Mizoguchi, Chizuko Watanabe, Akihiko Yonezawa, and Shinobu Sakurada

Regulated Exocytosis from Astrocytes: Physiological and Pathological Related Aspects
Corrado Calì, Julie Marchaland, Paola Spagnuolo, Julien Gremion, and Paola Bezzi

Glutamate Release from Astrocytic Gliosomes Under Physiological and Pathological Conditions
Marco Milanese, Tiziana Bonifacino, Simona Zappettini, Cesare Usai, Carlo Tacchetti, Mario Nobile, and Giambattista Bonanno

Neurotrophic and Neuroprotective Actions of an Enhancer of Ganglioside Biosynthesis
Jin-ichi Inokuchi

Involvement of Endocannabinoid Signaling in the Neuroprotective Effects of Subtype 1 Metabotropic Glutamate Receptor Antagonists in Models of Cerebral Ischemia
Elisa Landucci, Francesca Boscia, Elisabetta Gerace, Tania Scartabelli, Andrea Cozzi, Flavio Moroni, Guido Mannaioni, and Domenico E. Pellegrini-Giampietro

NF-kappaB Dimers in the Regulation of Neuronal Survival
Ilenia Sarnico, Annamaria Lanzillotta, Marina Benarese, Manuela Alghisi, Cristina Baiguera, Leontino Battistin, PierFranco Spano, and Marina Pizzi

Oxidative Stress in Stroke Pathophysiology: Validation of Hydrogen Peroxide Metabolism as a Pharmacological Target to Afford Neuroprotection
Diana Amantea, Maria Cristina Marrone, Robert Nisticò, Mauro Federici, Giacinto Bagetta, Giorgio Bernardi, and Nicola Biagio Mercuri

Role of Akt and ERK Signaling in the Neurogenesis following Brain Ischemia
Norifumi Shioda, Feng Han, and Kohji Fukunaga

Prevention of Glutamate Accumulation and Upregulation of Phospho-Akt may Account for Neuroprotection Afforded by Bergamot Essential Oil against Brain Injury Induced by Focal Cerebral Ischemia in Rat
Diana Amantea, Vincenza Fratto, Simona Maida, Domenicantonio Rotiroti, Salvatore Ragusa, Giuseppe Nappi, Giacinto Bagetta, and Maria Tiziana Corasaniti

Identification of Novel Pharmacological Targets to Minimize Excitotoxic Retinal Damage
Rossella Russo, Domenicantonio Rotiroti, Cristina Tassorelli, Carlo Nucci, Giacinto Bagetta, Massimo Gilberto Bucci, Maria Tiziana Corasaniti, and Luigi Antonio Morrone

INDEX

Volume 86

Section One: Hybrid Bionic Systems
EMG-Based and Gaze-Tracking-Based Man–Machine Interfaces
Federico Carpi and Danilo De Rossi

Bidirectional Interfaces with the Peripheral Nervous System
Silvestro Micera and Xavier Navarro

Interfacing Insect Brain for Space Applications
Giovanni Di Pino, Tobias Seidl, Antonella Benvenuto, Fabrizio Sergi, Domenico Campolo, Dino Accoto, Paolo Maria Rossini, and Eugenio Guglielmelli

Section Two: Meet the Brain
Meet the Brain: Neurophysiology
John Rothwell

Fundamentals of Electroencefalography, Magnetoencefalography, and Functional Magnetic Resonance Imaging
Claudio Babiloni, Vittorio Pizzella, Cosimo Del Gratta, Antonio Ferretti, and Gian Luca Romani

Implications of Brain Plasticity to Brain–Machine Interfaces Operation: A Potential Paradox?
Paolo Maria Rossini

Section Three: Brain Machine Interfaces, A New Brain-to-Environment Communication Channel
An Overview of BMIs
Francisco Sepulveda

Neurofeedback and Brain–Computer Interface: Clinical Applications
Niels Birbaumer, Ander Ramos Murguialday, Cornelia Weber, and Pedro Montoya

Flexibility and Practicality: Graz Brain–Computer Interface Approach
Reinhold Scherer, Gernot R. Müller-Putz, and Gert Pfurtscheller

On the Use of Brain–Computer Interfaces Outside Scientific Laboratories: Toward an Application in Domotic Environments
F. Babiloni, F. Cincotti, M. Marciani, S. Salinari, L. Astolfi, F. Aloise, F. De Vico Fallani, and D. Mattia

Brain–Computer Interface Research at the Wadsworth Center: Developments in Noninvasive Communication and Control
Dean J. Krusienski and Jonathan R. Wolpaw

Watching Brain TV and Playing Brain Ball: Exploring Novel BCL Strategies Using Real–Time Analysis of Human Intercranial Data
Karim Jerbi, Samson Freyermuth, Lorella Minotti, Philippe Kahane, Alain Berthoz, and Jean-Philippe Lachaux

Section Four: Brain-Machine Interfaces and Space
Adaptive Changes of Rhythmic EEG Oscillations in Space: Implications for Brain–Machine Interface Applications
G. Cheron, A. M. Cebolla, M. Petieau, A. Bengoetxea, E. Palmero-Soler, A. Leroy, and B. Dan

Validation of Brain–Machine Interfaces During Parabolic Flight
José del R. Millán, Pierre W. Ferrez, and Tobias Seidl

Matching Brain–Machine Interface Performance to Space Applications
Luca Citi, Oliver Tonet, and Martina Marinelli

Brain–Machine Interfaces for Space Applications—Research, Technological Development, and Opportunities
Leopold Summerer, Dario Izzo, and Luca Rossini

INDEX

Volume 87

Peripheral Nerve Repair and Regeneration Research: A Historical Note
Bruno Battiston, Igor Papalia, Pierluigi Tos, and Stefano Geuna

Development of the Peripheral Nerve
Suleyman Kaplan, Ersan Odaci, Bunyami Unal, Bunyamin Sahin, and Michele Fornaro

Histology of the Peripheral Nerve and Changes Occurring During Nerve Regeneration
Stefano Geuna, Stefania Raimondo, Giulia Ronchi, Federica Di Scipio, Pierluigi Tos, Krzysztof Czaja, and Michele Fornaro

Methods and Protocols in Peripheral Nerve Regeneration Experimental Research: Part I—Experimental Models
Pierluigi Tos, Giulia Ronchi, Igor Papalia, Vera Sallen, Josette Legagneux, Stefano Geuna, and Maria G. Giacobini-Robecchi

Methods and Protocols in Peripheral Nerve Regeneration Experimental Research: Part II—Morphological Techniques
Stefania Raimondo, Michele Fornaro, Federica Di Scipio, Giulia Ronchi, Maria G. Giacobini-Robecchi, and Stefano Geuna

Methods and Protocols in Peripheral Nerve Regeneration Experimental Research: Part III—Electrophysiological Evaluation
Xavier Navarro and Esther Udina

Methods and Protocols in Peripheral Nerve Regeneration Experimental Research: Part IV—Kinematic Gait Analysis to Quantify Peripheral Nerve Regeneration in the Rat
Luís M. Costa, Maria J. Simões, Ana C. Maurício and Artur S.P. Varejão

Current Techniques and Concepts in Peripheral Nerve Repair
Maria Siemionow and Grzegorz Brzezicki

Artificial Scaffolds for Peripheral Nerve Reconstruction
Valeria Chiono, Chiara Tonda-Turo, and Gianluca Ciardelli

Conduit Luminal Additives for Peripheral Nerve Repair
Hede Yan, Feng Zhang, Michael B. Chen, and William C. Lineaweaver

Tissue Engineering of Peripheral Nerves
Bruno Battiston, Stefania Raimondo, Pierluigi Tos, Valentina Gaidano, Chiara Audisio, Anna Scevola, Isabelle Perroteau, and Stefano Geuna

Mechanisms Underlying The End-to-Side Nerve Regeneration
Eleana Bontioti and Lars B. Dahlin

Experimental Results in End-To-Side Neurorrhaphy
Alexandros E. Beris and Marios G. Lykissas

End-to-Side Nerve Regeneration: From the Laboratory Bench to Clinical Applications
Pierluigi Tos, Stefano Artiaco, Igor Papalia, Ignazio Marcoccio, Stefano Geuna, and Bruno Battiston

Novel Pharmacological Approaches to Schwann Cells as Neuroprotective Agents for Peripheral Nerve Regeneration
Valerio Magnaghi, Patrizia Procacci, and Ada Maria Tata

Melatonin and Nerve Regeneration
Ersan Odaci and Suleyman Kaplan

Transthyretin: An Enhancer of Nerve Regeneration
Carolina E. Fleming, Fernando Milhazes Mar, Filipa Franquinho, and Mónica M. Sousa

Enhancement of Nerve Regeneration and Recovery by Immunosuppressive Agents
Damien P. Kuffler

The Role of Collagen in Peripheral Nerve Repair
Guido Koopmans, Birgit Hasse, and Nektarios Sinis

Gene Therapy Perspectives for Nerve Repair
Serena Zacchigna and Mauro Giacca

Use of Stem Cells for Improving Nerve Regeneration
Giorgio Terenghi, Mikael Wiberg, and Paul J. Kingham

Transplantation of Olfactory Ensheathing Cells for Peripheral Nerve Regeneration
Christine Radtke, Jeffery D. Kocsis, and Peter M. Vogt

Manual Stimulation of Target Muscles has Different Impact on Functional Recovery after Injury of Pure Motor or Mixed Nerves
Nektarios Sinis, Thodora Manoli, Frank Werdin, Armin Kraus, Hans E. Schaller, Orlando Guntinas-Lichius, Maria Grosheva, Andrey Irintchev, Emanouil Skouras, Sarah Dunlop, and Doychin N. Angelov

Electrical Stimulation for Improving Nerve Regeneration: Where do we Stand?
Tessa Gordon, Olewale A. R. Sulaiman, and Adil Ladak

Phototherapy in Peripheral Nerve Injury: Effects on Muscle Preservation and Nerve Regeneration
Shimon Rochkind, Stefano Geuna, and Asher Shainberg

Age-Related Differences in the Reinnervation after Peripheral Nerve Injury
Uroš Kovačič, Janez Sketelj, and Fajko F. Bajrović

Neural Plasticity After Nerve Injury and Regeneration
Xavier Navarro

Future Perspective in Peripheral Nerve Reconstruction
Lars Dahlin, Fredrik Johansson, Charlotta Lindwall, and Martin Kanje

INDEX

Volume 88

Effects Of Psychostimulants On Neurotrophins: Implications For Psychostimulant-Induced Neurotoxicity
Francesco Angelucci, Valerio Ricci, Gianfranco Spalletta, Carlo Caltagirone, Aleksander A. Mathé, and Pietro Bria

Dosing Time-Dependent Actions of Psychostimulants
H. Manev and T. Uz

Dopamine-Induced Behavioral Changes and Oxidative Stress in Methamphetamine-Induced Neurotoxicity
Taizo Kita, Ikuko Miyazaki, Masato Asanuma, Mika Takeshima, and George C. Wagner

Acute Methamphetamine Intoxication: Brain Hyperthermia, Blood–Brain Barrier, Brain Edema, and morphological cell abnormalities
Eugene A. Kiyatkin and Hari S. Sharma

Molecular Bases of Methamphetamine-Induced Neurodegeneration
Jean Lud Cadet and Irina N. Krasnova

Involvement of Nicotinic Receptors in Methamphetamine- and MDMA-Induced Neurotoxicity: Pharmacological Implications
E. Escubedo, J. Camarasa, C. Chipana, S. García-Ratés, and D.Pubill

Ethanol Alters the Physiology of Neuron–Glia Communication
Antonio González and Ginés M. Salido

Therapeutic Targeting of "DARPP-32": A Key Signaling Molecule in the Dopiminergic Pathway for the Treatment of Opiate Addiction
Supriya D. Mahajan, Ravikumar Aalinkeel, Jessica L. Reynolds, Bindukumar B. Nair, Donald E. Sykes, Zihua Hu, Adela Bonoiu, Hong Ding, Paras N. Prasad, and Stanley A. Schwartz

Pharmacological and Neurotoxicological Actions Mediated By Bupropion and Diethylpropion
Hugo R. Arias, Abel Santamaría, and Syed F. Ali

Neural and Cardiac Toxicities Associated With 3,4-Methylenedioxymethamphetamine (MDMA)
Michael H. Baumann and Richard B. Rothman

Cocaine-Induced Breakdown of the Blood–Brain Barrier and Neurotoxicity
Hari S. Sharma, Dafin Muresanu, Aruna Sharma, and Ranjana Patnaik

Cannabinoid Receptors in Brain: Pharmacogenetics, Neuropharmacology, Neurotoxicology, and Potential Therapeutic Applications
Emmanuel S. Onaivi

Intermittent Dopaminergic Stimulation causes Behavioral Sensitization in the Addicted Brain and Parkinsonism
Francesco Fornai, Francesca Biagioni, Federica Fulceri, Luigi Murri, Stefano Ruggieri, Antonio Paparelli

The Role of the Somatotrophic Axis in Neuroprotection and Neuroregeneration of the Addictive Brain
Fred Nyberg

INDEX

Volume 89

Molecular Profiling of Striatonigral and Striatopallidal Medium Spiny Neurons: Past, Present, and Future
Mary Kay Lobo

BAC to Degeneration: Bacterial Artificial Chromosome (Bac)-Mediated Transgenesis for Modeling Basal Ganglia Neurodegenerative Disorders
Xiao-Hong Lu

Behavioral Outcome Measures for the Assessment of Sensorimotor Function in Animal Models of Movement Disorders
Sheila M. Fleming

The Role of DNA Methylation in the Central Nervous System and Neuropsychiatric Disorders
Jian Feng and Guoping Fan

Heritability of Structural Brain Traits: An Endo-phenotype Approach to Deconstruct Schizophrenia
Nil Kaymaz and J. Van Os

The Role of Striatal NMDA Receptors in Drug Addiction
Yao-Ying Ma, Carlos Cepeda, and Cai-Lian Cui

Deciphering Rett Syndrome With Mouse Genetics, Epigenomics, and Human Neurons
Jifang Tao, Hao Wu, and Yi Eve Sun

INDEX

Volume 90

Part I: Introduction

Introductory Remarks on the History and Current Applications of TCS
Matthew B. Stern

Method and Validity of Transcranial Sonography in Movement Disorders
David Školoudík and Uwe Walter

Transcranial Sonography—Anatomy
Heiko Huber

Part II: Transcranial Sonography in Parkinsons Disease

Transcranial Sonography in Relation to SPECT and MIBG
Yoshinori Kajimoto, Hideto Miwa and Tomoyoshi Kondo

Diagnosis of Parkinson's Disease—Transcranial Sonography in Relation to MRI
Ludwig Niehaus and Kai Boelmans

Early Diagnosis of Parkinson's Disease
Alexandra Gaenslen and Daniela Berg

Transcranial Sonography in the Premotor Diagnosis of Parkinson's Disease
Stefanie Behnke, Ute Schroder and Daniela Berg

Pathophysiology of Transcranial Sonography Signal Changes in the Human Substantia Nigra
K. L. Double, G. Todd and S. R. Duma

Transcranial Sonography for the Discrimination of Idiopathic Parkinson's Disease from the Atypical Parkinsonian Syndromes
A. E. P. Bouwmans, A. M. M. Vlaar, K. Srulijes, W. H. Mess AND W. E. J. Weber

Transcranial Sonography in the Discrimination of Parkinson's Disease Versus Vascular Parkinsonism
Pablo Venegas-Francke

TCS in Monogenic Forms of Parkinson's Disease
Kathrin Brockmann and Johann Hagenah

Part III—Transcranial Sonography in other Movement Disorders and Depression

Transcranial Sonography in Brain Disorders with Trace Metal Accumulation
Uwe Walter

Transcranial Sonography in Dystonia
Alexandra Gaenslen

Transcranial Sonography in Essential Tremor
Heike Stockner and Isabel Wurster

VII—Transcranial Sonography in Restless Legs Syndrome
Jana Godau and Martin Sojer

Transcranial Sonography in Ataxia
Christos Krogias, Thomas Postert and Jens Eyding

Transcranial Sonography in Huntington's Disease
Christos Krogias, Jens Eyding and Thomas Postert

Transcranial Sonography in Depression
Milija D. Mijajlovic

Part IV: Future Applications and Conclusion

Transcranial Sonography-Assisted Stereotaxy and Follow-Up of Deep Brain Implants in Patients with Movement Disorders
Uwe Walter

Conclusions
Daniela Berg

INDEX

Volume 91

The Role of microRNAs in Drug Addiction: A Big Lesson from Tiny Molecules
Andrzej Zbigniew Pietrzykowski

The Genetics of Behavioral Alcohol Responses in *Drosophila*
Aylin R. Rodan and Adrian Rothenfluh

Neural Plasticity, Human Genetics, and Risk for Alcohol Dependence
Shirley Y. Hill

Using Expression Genetics to Study the Neurobiology of Ethanol and Alcoholism
Sean P. Farris, Aaron R. Wolen and Michael F. Miles

Genetic Variation and Brain Gene Expression in Rodent Models of Alcoholism: Implications for Medication Development
Karl Björk, Anita C. Hansson and W. olfgang H. Sommer

Identifying Quantitative Trait Loci (QTLs) and Genes (QTGs) for Alcohol-Related Phenotypes in Mice
Lauren C. Milner and Kari J. Buck

Glutamate Plasticity in the Drunken Amygdala: The Making of an Anxious Synapse
Brian A. Mccool, Daniel T. Christian, Marvin R. Diaz and Anna K. Läck

Ethanol Action on Dopaminergic Neurons in the Ventral Tegmental Area: Interaction with Intrinsic Ion Channels and Neurotransmitter Inputs
Hitoshi Morikawa and Richard A. Morrisett

Alcohol and the Prefrontal Cortex
Kenneth Abernathy, L. Judson Chandler and John J. Woodward

BK Channel and Alcohol, A Complicated Affair
Gilles Erwan Martin

A Review of Synaptic Plasticity at Purkinje Neurons with a Focus on Ethanol-Induced Cerebellar Dysfunction
C. Fernando Valenzuela, Britta Lindquist and Paula A. Zflmudio-Bulcock

INDEX

Volume 92

The Development of the Science of Dreaming
Claude Gottesmann

Dreaming as Inspiration: Evidence from Religion, Philosophy, Literature, and Film
Kelly Bulkeley

Developmental Perspective: Dreaming Across the Lifespan and What This Tells Us
Melissa M. Burnham and Christian Conte

REM and NREM Sleep Mentation
Patrick Mcnamara, Patricia Johnson, Deirdre McLaren, Erica Harris,Catherine Beauharnais and Sanford Auerbach

Neuroimaging of Dreaming: State of the Art and Limitations
Caroline Kussé, Vincenzo Muto, Laura Mascetti, Luca Matarazzo, Ariane Foret, Anahita Shaffii-Le Bourdiec and Pierre Maquet

Memory Consolidation, The Diurnal Rhythm of Cortisol, and The Nature of Dreams: A New Hypothesis
Jessica D. Payne

Characteristics and Contents of Dreams
Michael Schredl

Trait and Neurobiological Correlates of Individual Differences in Dream Recall and Dream Content
Mark Blagrove and Edward F. Pace-Schott

Consciousness in Dreams
David Kahn and Tzivia Gover

The Underlying Emotion and the Dream: Relating Dream Imagery to the Dreamer's Underlying Emotion can Help Elucidate the Nature of Dreaming
Ernest Hartmann

Dreaming, Handedness, and Sleep Architecture: Interhemispheric Mechanisms
Stephen D. Christman and Ruth E. Propper

To What Extent Do Neurobiological Sleep-Waking Processes Support Psychoanalysis?
Claude Gottesmann

The Use of Dreams in Modern Psychotherapy
Clara E. Hill and Sarah Knox

INDEX

Volume 93

Underlying Brain Mechanisms that Regulate Sleep-Wakefulness Cycles
Irma Gvilia

What Keeps Us Awake?—the Role of Clocks and Hourglasses, Light, and Melatonin
Christian Cajochen, Sarah Chellappa and Christina Schmidt

Suprachiasmatic Nucleus and Autonomic Nervous System Influences on Awakening From Sleep
Andries Kalsbeek, Chun-xia Yi, Susanne E. la Fleur, Ruud m. Buijs, and Eric Fliers

Preparation for Awakening: Self-Awakening Vs. Forced Awakening: Preparatory Changes in the Pre-Awakening Period
Mitsuo Hayashi, Noriko Matsuura and Hiroki Ikeda

Circadian and Sleep Episode Duration Influences on Cognitive Performance Following the Process of Awakening
Robert L. Matchock

The Cortisol Awakening Response in Context
Angela Clow, Frank Hucklebridge and Lisa Thorn

Causes and Correlates of Frequent Night Awakenings in Early Childhood
Amy Jo Schwichtenberg and Beth Goodlin-Jones

Pathologies of Awakenings: The Clinical Problem of Insomnia Considered From Multiple Theory Levels
Douglas E. Moul

The Neurochemistry of Awakening: Findings from Sleep Disorder Narcolepsy
Seiji Nishino and Yohei Sagawa

INDEX

Volume 94

5-HT_6 Medicinal Chemistry
Kevin G. Liu and Albert J. Robichaud

Patents
Nicolas Vincent Ruiz and Gloria Oranias

5-HT_6 Receptor Charactertization
Teresa Riccioni

5-HT_6 Receptor Signal Transduction: Second Messenger Systems
Xavier Codony, Javier Burgueño, Maria Javier Ramírez and José Miguel Vela

Electrophysiology of 5-HT_6 Receptors
Annalisa Tassone, Graziella Madeo, Giuseppe Sciamanna, Antonio Pisani and Paola Bonsi

Genetic Variations and Association
Massimo Gennarelli and Annamaria Cattaneo

Pharmacokinetics of 5-HT_6 Receptor Ligands
Angelo Mancinelli

INDEX

Volume 95

Introductory Remarks: Catechol-O-Methyltransferase Inhibition–An Innovative Approach to Enhance L-dopa Therapy in Parkinson's Disease with Dual Enzyme Inhibition
Erkki Nissinen

The Catechol-O-Methyltransferase Gene: its Regulation and Polymorphisms
Elizabeth M. Tunbridge

Distribution and Functions of Catechol-O-Methyltransferase Proteins: Do Recent Findings Change the Picture?
Timo T. Myöhänen and Pekka T. Männistö

Catechol-O-Methyltransferase Enzyme: Cofactor S-Adenosyl-L-Methionine and Related Mechanisms
Thomas Müller

Biochemistry and Pharmacology of Catechol-O-Methyltransferase Inhibitors
Erkki nissinen and Pekka T. Männisto

The Chemistry of Catechol-O-Methyltransferase Inhibitors
David A. Learmonth, László E. Kiss, and Patrício Soares-da-Silva

Toxicology and Safety of COMT Inhibitors
Kristiina Haasio

Catechol-O-Methyltransferase Inhibitors in Preclinical Models as Adjuncts of L-dopa Treatment
Concepció Marin and J. A. Obeso

Problems with the Present Inhibitors and a Relevance of New and Improved COMT Inhibitors in Parkinson's Disease
Seppo Kaakkola

Catechol-O-Methyltransferase and Pain
Oleg Kambur and Pekka T. Männistö

INDEX

Volume 96

The Central Role of 5-HT_6 Receptors in Modulating Brain Neurochemistry
Lee A. Dawson

5-HT_6 Receptor Memory and Amnesia: Behavioral Pharmacology – Learning and Memory Processes
Alfredo Meneses, G. Pérez-García, R. Tellez, T. Ponce-Lopez and C. Castillo

Behavioral Pharmacology: Potential Antidepressant and Anxiolytic Properties
Anna Wesołowska and Magdalena Jastrzbska-Wisek

The 5-HT_6 Receptor as a Target for Developing Novel Antiobesity Drugs
David Heal, Jane Gosden and Sharon Smith

Behavioral and Neurochemical Pharmacology of 5-HT_6 Receptors Related to Reward and Reinforcement
Gaetano Di Chiara, Valentina Valentini and Sandro Fenu

5-HT_6 Receptor Ligands and their Antipsychotic Potential
Jørn Arnt and Christina Kurre Olsen

5-HT_6 Receptor Ligands as Antidementia Drugs
Ellen Siobhan Mitchell

Other 5-HT_6 Receptor-Mediated Effects
Franco Borsini

INDEX

Volume 97

Behavioral Pharmacology of Orofacial Movement Disorders
Noriaki Koshikawa, Satoshi Fujita and Kazunori Adachi

Regulation of Orofacial Movement: Dopamine Receptor Mechanisms and Mutant Models
John L. Waddington, Gerard J. O'Sullivan and Katsunori Tomiyama

Regulation of Orofacial Movement: Amino Acid Mechanisms and Mutant Models
Katsunori Tomiyama, Colm M.P. O'Tuathaigh, and John L. Waddington

The Trigeminal Circuits Responsible for Chewing
Karl-Gunnar Westberg and Arlette Kolta

Ultrastructural Basis for Craniofacial Sensory Processing in the Brainstem
Yong Chul Bae and Atsushi Yoshida

Mechanisms of Nociceptive Transduction and Transmission: A Machinery for Pain Sensation and Tools for Selective Analgesia
Alexander M. Binshtok

Peripheral and Central Mechanisms of Orofacial Inflammatory Pain
Barry J. Sessle

The Role of Trigeminal Interpolaris-Caudalis Transition Zone in Persistent Orofacial Pain
Ke Ren and Ronald Dubner

Physiological Mechanisms of Neuropathic Pain: The Orofacial Region
Koichi Iwata, Yoshiki Imamura, Kuniya Honda and Masamichi Shinoda

Neurobiology of Estrogen Status in Deep Craniofacial Pain
David A Bereiter and Keiichiro Okamoto

Macroscopic Connection of Rat Insular Cortex: Anatomical Bases Underlying its Physiological Functions
Masayuki Kobayashi

The Balance Between Excitation And Inhibition And Functional Sensory Processing in the Somatosensory Cortex
Zhi Zhang and Qian-Quan Sun

INDEX

Volume 98

An Introduction to Dyskinesia—the Clinical Spectrum
Ainhi Ha and Joseph Jankovic

L-dopa-induced Dyskinesia—Clinical Presentation, Genetics, And Treatment
L.K. Prashanth, Susan Fox and Wassilios G. Meissner

Experimental Models of L-DOPA-induced Dyskinesia
Tom H. Johnston and Emma L. Lane

Molecular Mechanisms of L-DOPA-induced Dyskinesia
Gilberto Fisone and Erwan Bezard

New Approaches to Therapy
Jonathan Brotchie and Peter Jenner

Surgical Approach to L-DOPA-induced Dyskinesias
Tejas Sankar and Andres M. Lozano

Clinical and Experimental Experiences of Graft-induced Dyskinesia
Emma L. Lane

Tardive Dyskinesia: Clinical Presentation and Treatment
P.N. van Harten and D.E. Tenback

Epidemiology and Risk Factors for (Tardive) Dyskinesia
D.E. Tenback and P.N. van Harten

Genetics of Tardive Dyskinesia
Heon-Jeong Lee and Seung-Gul Kang

Animal Models of Tardive Dyskinesia
S.K. Kulkarni and Ashish Dhir

Surgery for Tardive Dyskinesia
Stephane Thobois, Alice Poisson and Philippe Damier

Huntington's Disease: Clinical Presentation and Treatment
M.J.U. Novak and S.J. Tabrizi

Genetics and Neuropathology of Huntington's Disease: Huntington's Disease
Anton Reiner, Ioannis Dragatsis and Paula Dietrich

Pathogenic Mechanisms in Huntington's Disease
Lesley Jones and Alis Hughes

Experimental Models of HD And Reflection on Therapeutic Strategies
Olivia L. Bordiuk, Jinho Kim and Robert J. Ferrante

Cell-based Treatments for Huntington's Disease
Stephen B. Dunnett and Anne E. Rosser

Clinical Phenomenology of Dystonia
Carlo Colosimo and Alfredo Berardelli

Genetics and Pharmacological Treatment of Dystonia
Susan Bressman and Matthew James

Experimental Models of Dystonia
A. Tassone, G. Sciamanna, P. Bonsi, G. Martella and A. Pisani

Surgical Treatment of Dystonia
John Yianni, Alexander L. Green and Tipu Z. Aziz

INDEX

Volume 99

Seizure and Epilepsy: Studies of Seizure-disorders in *Drosophila*
Louise Parker, Iris C. Howlett, Zeid M. Rusan and Mark A. Tanouye

Homeostatic Control of Neural Activity: A *Drosophila* Model for Drug Tolerance and Dependence
Alfredo Ghezzi and Nigel S. Atkinson

Attention in *Drosophila*
Bruno van Swinderen

The roles of *Fruitless* and *Doublesex* in the Control of Male Courtship
Brigitte Dauwalder

Circadian Plasticity: from Structure to Behavior
Lia Frenkel and María Fernanda Ceriani

Learning and Memory in *Drosophila*: Behavior, Genetics, and Neural Systems
Lily Kahsai and Troy Zars

Studying Sensorimotor Processing with Physiology in Behaving *Drosophila*
Johannes D. Seelig and Vivek Jayaraman

Modeling Human Trinucleotide Repeat Diseases in *Drosophila*
Zhenming Yu and Nancy M. Bonini

From Genetics to Structure to Function: Exploring Sleep in *Drosophila*
Daniel Bushey and Chiara Cirelli

INDEX

Volume 100

Structural Properties of Human Monoamine Oxidases A and B
Claudia Binda, Andrea Mattevi and Dale E. Edmondson

Behavioral Outcomes of Monoamine Oxidase Deficiency: Preclinical and Clinical Evidence
Marco Bortolato and Jean C. Shih

Kinetic Behavior and Reversible Inhibition of Monoamine Oxidases—Enzymes that Many Want Dead
Keith F. Tipton, Gavin P. Davey and Andrew G. McDonald

The Pharmacology of Selegiline
Kálmán Magyar

Type A Monoamine Oxidase Regulates Life and Death of Neurons in Neurodegeneration and Neuroprotection
Makoto Naoi, Wakako Maruyama, Keiko Inaba-Hasegawa and Yukihiro Akao

Multimodal Drugs and their Future for Alzheimer's and Parkinson's Disease
Cornelis J. Van der Schyf and Werner J. Geldenhuys

Neuroprotective Profile of the Multitarget Drug Rasagiline in Parkinson's Disease
Orly Weinreb, Tamar Amit, Peter Riederer, Moussa B.H. Youdim and Silvia A. Mandel

Rasagiline in Parkinson's Disease
L.M. Chahine and M.B. Stern

Selective Inhibitors of Monoamine Oxidase Type B and the "Cheese Effect"
John P.M. Finberg and Ken Gillman

A Novel Anti-Alzheimer's Disease Drug, Ladostigil: Neuroprotective, Multimodal Brain-Selective Monoamine Oxidase and Cholinesterase Inhibitor
Orly Weinreb, Tamar Amit, Orit Bar-Am and Moussa B.H. Youdim

Novel MAO-B Inhibitors: Potential Therapeutic Use of the Selective MAO-B Inhibitor PF9601N in Parkinson's Disease
Mercedes Unzeta and Elisenda Sanz

INDEX

Volume 101

General Overview: Biomarkers in Neuroscience Research
Michaela D. Filiou and Christoph W. Turck

Imaging Brain Microglial Activation Using Positron Emission Tomography and Translocator Protein-Specific Radioligands
David R.J. Owen and Paul M. Matthews

The Utility of Gene Expression in Blood Cells for Diagnosing Neuropsychiatric Disorders
Christopher H. Woelk, Akul Singhania, Josué Pérez-Santiago, Stephen J. Glatt and Ming T. Tsuang

Proteomic Technologies for Biomarker Studies in Psychiatry: Advances and Needs
Daniel Martins-de-Souza, Paul C. Guest, Natacha Vanattou-Saifoudine, Laura W. Harris and Sabine Bahn

Converging Evidence of Blood-Based Biomarkers for Schizophrenia: An update
Man K. Chan, Paul C. Guest, Yishai Levin, Yagnesh Umrania, Emanuel Schwarz, Sabine Bahn and Hassan Rahmoune

Abnormalities in Metabolism and Hypothalamic–Pituitary–Adrenal Axis Function in Schizophrenia
Paul C. Guest, Daniel Martins-de-Souza, Natacha Vanattou-Saifoudine, Laura W. Harris and Sabine Bahn

Immune and Neuroimmune Alterations in Mood Disorders and Schizophrenia
Roosmarijn C. Drexhage, Karin Weigelt, Nico van Beveren, Dan Cohen, Marjan A. Versnel, Willem A. Nolen and Hemmo A. Drexhage

Behavioral and Molecular Biomarkers in Translational Animal Models for Neuropsychiatric Disorders
Zoltán Sarnyai, Murtada Alsaif, Sabine Bahn, Agnes Ernst, Paul C. Guest, Eva Hradetzky, Wolfgang Kluge, Viktoria Stelzhammer and Hendrik Wesseling

Stem Cell Models for Biomarker Discovery in Brain Disease
Alan Mackay-Sim, George Mellick and Stephen Wood

The Application of Multiplexed Assay Systems for Molecular Diagnostics
Emanuel Schwarz, Nico J.M. VanBeveren, Paul C. Guest, Rauf Izmailov and Sabine Bahn

Algorithm Development for Diagnostic Biomarker Assays
Rauf Izmailov, Paul C. Guest, Sabine Bahn and Emanuel Schwarz

Challenges of Introducing New Biomarker Products for Neuropsychiatric Disorders into the Market
Sabine Bahn, Richard Noll, Anthony Barnes, Emanuel Schwarz and Paul C. Guest

Toward Personalized Medicine in the Neuropsychiatric Field
Erik H.F. Wong, Jayne C. Fox, Mandy Y.M. Ng and Chi-Ming Lee

Clinical Utility of Serum Biomarkers for Major Psychiatric Disorders
Nico J.M. van Beveren and Witte J.G. Hoogendijk

The Future: Biomarkers, Biosensors, Neuroinformatics, and E-Neuropsychiatry
Christopher R. Lowe

SUBJECT INDEX

Volume 102

The Function and Mechanisms of Nurr1 Action in Midbrain Dopaminergic Neurons, from Development and Maintenance to Survival
Yu Luo

Monoclonal Antibodies as Novel Neurotherapeutic Agents in CNS Injury and Repair
Aruna Sharma and Hari Shanker Sharma

The Blood–Brain Barrier in Alzheimer's Disease: Novel Therapeutic Targets and Nanodrug delivery
Hari Shanker Sharma, Rudy J. Castellani, Mark A. Smith and Aruna Sharma

Neurovascular Aspects of Amyotrophic Lateral Sclerosis
Maria Carolina O. Rodrigues, Diana G. Hernandez-Ontiveros, Michael K. Louis, Alison E. Willing, Cesario V. Borlongan, Paul R. Sanberg, Júlio C. Voltarelli and Svitlana Garbuzova-Davis

Quercetin in Hypoxia-Induced Oxidative Stress: Novel Target for Neuroprotection
Anand Kumar Pandey, Ranjana Patnaik, Dafin F. Muresanu, Aruna Sharma and Hari Shanker Sharma

Environmental Conditions Modulate Neurotoxic Effects of Psychomotor Stimulant Drugs of Abuse
Eugene A. Kiyatkin and Hari Shanker Sharma

Central Nervous Tissue Damage after Hypoxia and Reperfusion in Conjunction with Cardiac Arrest and Cardiopulmonary Resuscitation: Mechanisms of Action and Possibilities for Mitigation
Lars Wiklund, Cecile Martijn, Adriana Miclescu, Egidijus Semenas, Sten Rubertsson and Hari Shanker Sharma

Interactions Between Opioids and Anabolic Androgenic Steroids: Implications for the Development of Addictive Behavior
Fred Nyberg and Mathias Hallberg

Neurotrophic Factors and Neurodegenerative Diseases: A Delivery Issue
Barbara Ruozi, Daniela Belletti, Lucia Bondioli, Alessandro De Vita, Flavio Forni, Maria Angela Vandelli and Giovanni Tosi

Neuroprotective Effects of Cerebrolysin, a Combination of Different Active Fragments of Neurotrophic Factors and Peptides on the Whole Body Hyperthermia-Induced Neurotoxicity: Modulatory Roles of Co-morbidity Factors and Nanoparticle Intoxication
Hari Shanker Sharma, Aruna Sharma, Herbert Mössler and Dafin Fior Muresanu

Alzheimer's Disease and Amyloid: Culprit or Coincidence?
Stephen D. Skaper

Vascular Endothelial Growth Factor and Other Angioglioneurins: Key Molecules in Brain Development and Restoration
José Vicente Lafuente, Naiara Ortuzar, Harkaitz Bengoetxea, Susana Bulnes and Enrike G. Argandoña

INDEX

Volume 103

Lost and Found in Behavioral Informatics
Melissa A. Haendel and Elissa J. Chesler

Biological Databases for Behavioral Neurobiology
Erich J. Baker

A Survey of the Neuroscience Resource Landscape: Perspectives from the Neuroscience Information Framework
Jonathan Cachat, Anita Bandrowski, Jeffery S. Grethe, Amarnath Gupta, Vadim Astakhov, Fahim Imam, Stephen D. Larson, and Maryann E. Martone

The Neurobehavior Ontology: An Ontology for Annotation and Integration of Behavior and Behavioral Phenotypes
Georgios V. Gkoutos, Paul N. Schofield, and Robert Hoehndorf

Ontologies for Human Behavior Analysis and Their Application to Clinical Data
Janna Hastings and Stefan Schulz

Text-Mining and Neuroscience
Kyle H. Ambert and Aaron M. Cohen

Applying *In Silico* Integrative Genomics to Genetic Studies of Human Disease: A Review
Scott F. Saccone

SUBJECT INDEX

Volume 104

Cross Species Integration of Functional Genomics Experiments
Jeremy J. Jay

Model Organism Databases in Behavioral Neuroscience
Mary Shimoyama, Jennifer R. Smith, G. Thomas Hayman, Victoria Petri, and Rajni Nigam

Accessing and Mining Data from Large-Scale Mouse Phenotyping Projects
Hugh Morgan, Michelle Simon, and Ann-Marie Mallon

Bioinformatics Resources for Behavior Studies in the Laboratory Mouse
Carol J. Bult

Using Genome-Wide Expression Profiling to Define Gene Networks Relevant to the Study of Complex Traits: From RNA Integrity to Network Topology
M.A. O'Brien, B.N. Costin, and M.F. Miles

Genetic and Molecular Network Analysis of Behavior
Robert W. Williams and Megan K. Mulligan

Large-Scale Neuroinformatics for *In Situ* Hybridization Data in the Mouse Brain
Lydia L. Ng, Susan M. Sunkin, David Feng, Chris Lau, Chinh Dang, and Michael J. Hawrylycz

Opportunities for Bioinformatics in the Classification of Behavior and Psychiatric Disorders
Elissa J. Chesler and Ryan W. Logan

SUBJECT INDEX

Volume 105

Optic Nerve Disease and Axon Pathophysiology
Alireza Ghaffarieh and Leonard A. Levin

Role of Electrical Activity of Neurons for Neuroprotection
Takeshi Morimoto

Molecular Control of Axon Growth: Insights from Comparative Gene Profiling and High-Throughput Screening
Murray G. Blackmore

Gatekeeper Between Quiescence and Differentiation: p53 in Axonal Outgrowth and Neurogenesis
Giorgia Quadrato and Simone Di Giovanni

Cyclin-Dependent Kinase 5 in Axon Growth and Regeneration
Tao Ye, Amy K. Y. Fu, and Nancy Y. Ip

Rho Signaling and Axon Regeneration
L. McKerracher, Gino B. Ferraro, and Alyson E. Fournier

Neuron-Intrinsic Inhibitors of Axon Regeneration: PTEN and SOCS3
Xueting Luo and Kevin K. Park

INDEX

Volume 106

Neurotrophic Factors and the Regeneration of Adult Retinal Ganglion Cell Axons
Alan R. Harvey, Jacob Wei Wei Ooi, and Jennifer Rodger

MBS: Signaling Endosomes and Growth Cone Motility in Axon Regeneration
Michael B. Steketee and Jeffrey L. Goldberg

Intrinsic Mechanisms Regulating Axon Regeneration: An Integrin Perspective
Richard Eva, Melissa R. Andrews, Elske H.P. Franssen, and James W. Fawcett

The Role of Serotonin in Axon and Dendrite Growth
Ephraim F. Trakhtenberg and Jeffrey L. Goldberg

Inflammatory Pathways in Spinal Cord Injury
Samuel David, Juan Guillermo Zarruk, and Nader Ghasemlou

Combinatorial Therapy Stimulates Long-Distance Regeneration, Target Reinnervation, and Partial Recovery of Vision After Optic Nerve Injury in Mice
Silmara de Lima, Ghaith Habboub, and Larry I. Benowitz

From Bench to Beside to Cure Spinal Cord Injury: Lost in Translation?
Andreas Hug and Norbert Weidner

SUBJECT INDEX

Volume 107

Neuromodulation: A More Comprehensive Concept Beyond Deep Brain Stimulation
Clement Hamani and Elena Moro

Computational Models of Neuromodulation
Christopher R. Butson

Neurophysiology of Deep Brain Stimulation
Manuela Rosa, Gaia Giannicola, Sara Marceglia, Manuela Fumagalli, Sergio Barbieri, and Alberto Priori

Neurophysiology of Cortical Stimulation
Jean-Pascal Lefaucheur

Neural Mechanisms of Spinal Cord Stimulation
Robert D. Foreman and Bengt Linderoth

Magnetoencephalography and Neuromodulation
Alfons Schnitzler and Jan Hirschmann

Current Challenges to the Clinical Translation of Brain Machine Interface Technology
Charles W. Lu, Parag G. Patil, and Cynthia A. Chestek

Nanotechnology in Neuromodulation
Russell J. Andrews

Optogenetic Neuromodulation
Paul S. A. Kalanithi and Jaimie M. Henderson

Diffusion Tensor Imaging and Neuromodulation: DTI as Key Technology for Deep Brain Stimulation
Volker Arnd Coenen, Thomas E. Schlaepfer, Niels Allert, and Burkhard Mädler

DBS and Electrical Neuro-Network Modulation to Treat Neurological Disorders
Amanda Thompson, Takashi Morishita, and Michael S. Okun

Neuromodulation in Psychiatric Disorders
Yasin Temel, Sarah A. Hescham, Ali Jahanshahi, Marcus L. F. Janssen, Sonny K. H. Tan, Jacobus J. van Overbeeke, Linda Ackermans, Mayke Oosterloo, Annelien Duits, Albert F. G. Leentjens, and LeeWei Lim

Ethical Aspects of Neuromodulation
Christiane Woopen

Subject Index

Volume 108

Tissue Engineering and Regenerative Medicine: Past, Present, and Future
António J. Salgado, Joaquim M. Oliveira, Albino Martins, Fábio G. Teixeira, Nuno A. Silva, Nuno M. Neves, Nuno Sousa, and Rui L. Reis

Tissue Engineering and Peripheral Nerve Reconstruction: An Overview
Stefano Geuna, S. Gnavi, I. Perroteau, Pierluigi Tos, and B. Battiston

Bone Marrow Mesenchymal Stem Cell Transplantation for Improving Nerve Regeneration
Júlia Teixeira Oliveira, Klauss Mostacada, Silmara de Lima, and Ana Maria Blanco Martinez

Perspectives of Employing Mesenchymal Stem Cells from the Wharton's Jelly of the Umbilical Cord for Peripheral Nerve Repair
Jorge Ribeiro, Andrea Gartner, Tiago Pereira, Raquel Gomes, Maria Ascensão Lopes, Carolina Gonçalves, Artur Varejão, Ana Lúcia Luís, and Ana Colette Maurício

Adipose-Derived Stem Cells and Nerve Regeneration: Promises and Pitfalls
Alessandro Faroni, Giorgio Terenghi, and Adam J. Reid

The Pros and Cons of Growth Factors and Cytokines in Peripheral Axon Regeneration
Lars Klimaschewski, Barbara Hausott, and Doychin N. Angelov

Role of Inflammation and Cytokines in Peripheral Nerve Regeneration
P. Dubový, R. Jančálek, and T. Kubek

Ghrelin: A Novel Neuromuscular Recovery Promoting Factor?
Raimondo Stefania, Ronchi Giulia, Geuna Stefano, Pascal Davide, Reano Simone, Filigheddu Nicoletta, and Graziani Andrea

Neuregulin 1 Role in Schwann Cell Regulation and Potential Applications to Promote Peripheral Nerve Regeneration
Giovanna Gambarotta, Federica Fregnan, Sara Gnavi, and Isabelle Perroteau

Extracellular Matrix Components in Peripheral Nerve Regeneration
Francisco Gonzalez-Perez, Esther Udina, and Xavier Navarro

Subject Index

Volume 109

The Use of Chitosan-Based Scaffold to Enhance Regeneration in the Nervous System
Sara Gnavi, Christina Barwig, Thomas Freier, Kirsten Haarstert-Talini, Claudia Grothe, and Stefano Geuna

Interfaces with the Peripheral Nerve for the Control of Neuroprostheses
Jaume del Valle and Xavier Navarro

The Use of Shock Waves in Peripheral Nerve Regeneration: New Perspectives?
Thomas Hausner and Antal Nógrádi

Phototherapy and Nerve Injury: Focus on Muscle Response
Shimon Rochkind, Stefano Geuna, and Asher Shainberg

Electrical Stimulation for Promoting Peripheral Nerve Regeneration
Kirsten Haastert-Talini and Claudia Grothe

Role of Physical Exercise for Improving Post-traumatic Nerve Regeneration
Paulo A.S. Armada-da-Silva, Cátia Pereira, SandraAmado, and António P. Veloso

The Role of Timing in Nerve Reconstruction
Lars B. Dahlin

Future Perspectives in Nerve Repair and Regeneration
Pierluigi Tos, Giulia Ronchi, Stefano Geuna, and Bruno Battiston

INDEX

Volume 110

The Relevance of Metals in the Pathophysiology of Neurodegeneration, Pathological Considerations
Kurt A. Jellinger

Pantothenate Kinase-Associated Neurodegeneration (PKAN) and *PLA2G6*-Associated Neurodegeneration (PLAN): Review of Two Major Neurodegeneration with Brain Iron Accumulation (NBIA) Phenotypes
Manju A. Kurian and Susan J. Hayflick

Mitochondrial Membrane Protein-Associated Neurodegeneration (MPAN)
Monika Hartig, Holger Prokisch, Thomas Meitinger, and Thomas Klopstock

BPAN: The Only X-Linked Dominant NBIA Disorder
T.B. Haack, P. Hogarth, A. Gregory, P. Prokisch, and S.J. Hayflick

Neuroferritinopathy
M.J. Keogh, C.M. Morris, and P.F. Chinnery

Aceruloplasminemia: An Update
Satoshi Kono

Therapeutic Advances in Neurodegeneration with Brain Iron Accumulation
Giovanna Zorzi and Nardo Nardocci

The Neuropathology of Neurodegeneration with Brain Iron Accumulation
Michael C. Kruer

Imaging of Iron
Petr Dusek, Monika Dezortova, and Jens Wuerfel

The Role of Iron Imaging in Huntington's Disease
S.J.A. van den Bogaard, E.M. Dumas, and R.A.C. Roos

Lysosomal Storage Disorders and Iron
Jose Miguel Bras

Manganese and the Brain
Karin Tuschl, Philippa B. Mills, and Peter T. Clayton

Update on Wilson Disease
Aggarwal Annu and Bhatt Mohit

An Update on Primary Familial Brain Calcification
R.R. Lemos, J.B.M.M. Ferreira, M.P. Keasey, and J.R.M. Oliveira

INDEX

Volume 111

History of Acupuncture Research
Yi Zhuang, Jing-jing Xing, Juan Li, Bai-Yun Zeng, and Fan-rong Liang

Effects of Acupuncture Needling with Specific Sensation on Cerebral Hemodynamics and Autonomic Nervous Activity in Humans
Kouich Takamoto, Susumu Urakawa, Kazushige Sakai, Taketoshi Ono, and Hisao Nishijo

Acupuncture Point Specificity
Jing-jing Xing, Bai-Yun Zeng, Juan Li, Yi Zhuang, and Fan-rong Liang

Acupuncture Stimulation Induces Neurogenesis in Adult Brain
Min-Ho Nam, Kwang Seok Ahn, and Seung-Hoon Choi

Acupuncture and Neurotrophin Modulation
Marzia Soligo, Stefania Lucia Nori, Virginia Protto, Fulvio Florenzano, and Luigi Manni

Acupuncture Stimulation and Neuroendocrine Regulation
Jung-Sheng Yu, Bai-Yun Zeng, and Ching-Liang Hsieh

Current Development of Acupuncture Research in Parkinson's Disease
Bai-Yun Zeng, Sarah Salvage, and Peter Jenner

Acupuncture Therapy for Stroke Patients
Xin Li and Qiang Wang

Effects of Acupuncture Therapy on Alzheimer's Disease
Bai-Yun Zeng, Sarah Salvage, and Peter Jenner

Acupuncture Therapy for Psychiatric Illness
Karen Pilkington

Acupuncture for the Treatment of Insomnia
Kaicun Zhao

Acupuncture for the Treatment of Drug Addiction
Cai-Lian Cui, Liu-Zhen Wu, and Yi-jing Li

Acupuncture Regulation of Blood Pressure: Two Decades of Research
John C. Longhurst and Stephanie Tjen-A-Looi

Effect and Mechanism of Acupuncture on Gastrointestinal Diseases
Toku Takahashi

INDEX

Volume 112

An Introduction to the Clinical Phenomenology of Tourette Syndrome
Davide Martino, Namrata Madhusudan, Panagiotis Zis, and Andrea E. Cavanna

Functional Neuroanatomy of Tics
Irene Neuner, Frank Schneider, and N. Jon Shah

Functional Imaging of Dopaminergic Neurotransmission in Tourette Syndrome
Bàrbara Segura and Antonio P. Strafella

Nondopaminergic Neurotransmission in the Pathophysiology of Tourette Syndrome
Patrick T. Udvardi, Ester Nespoli, Francesca Rizzo, Bastian Hengerer, and Andrea G. Ludolph

Reinforcement Learning and Tourette Syndrome
Stefano Palminteri and Mathias Pessiglione

Genetic Susceptibility and Neurotransmitters in Tourette Syndrome
Peristera Paschou, Thomas V. Fernandez, Frank Sharp, Gary A. Heiman, and Pieter J. Hoekstra

Pharmacological Animal Models of Tic Disorders
Kevin W. McCairn and Masaki Isoda

Animal Models Recapitulating the Multifactorial Origin of Tourette Syndrome
Simone Macrì, Martina Proietti Onori, Veit Roessner, and Giovanni Laviola

Neuroendocrine Aspects of Tourette Syndrome
Davide Martino, Antonella Macerollo, and James F. Leckman

Clinical Pharmacology of Dopamine-Modulating Agents in Tourette's Syndrome
Sabine Mogwitz, Judith Buse, Stefan Ehrlich, and Veit Roessner

Clinical Pharmacology of Nondopaminergic Drugs in Tourette Syndrome
Andreas Hartmann

Antiepileptic Drugs and Tourette Syndrome
Andrea E. Cavanna and Andrea Nani

Clinical Pharmacology of Comorbid Obsessive–Compulsive Disorder in Tourette Syndrome
Valeria Neri and Francesco Cardona

Clinical Pharmacology of Comorbid Attention Deficit Hyperactivity Disorder in Tourette Syndrome
Renata Rizzo and Mariangela Gulisano

Emerging Treatment Strategies in Tourette Syndrome: What's in the Pipeline?
C. Termine, C. Selvini, G. Rossi, and U. Balottin

Tics and Other Stereotyped Movements as Side Effects of Pharmacological Treatment
Marcos Madruga-Garrido and Pablo Mir

INDEX

Volume 113

Autism Spectrum Disorder and the Cerebellum
Esther B.E. Becker and Catherine J. Stoodley

Contribution of Long Noncoding RNAs to Autism Spectrum Disorder Risk
Brent Wilkinson and Daniel B. Campbell

Identifying Essential Cell Types and Circuits in Autism Spectrum Disorders
Susan E. Maloney, Michael A. Rieger, and Joseph D. Dougherty

Connecting Signaling Pathways Underlying Communication to ASD Vulnerability
Stephanie Lepp, Ashley Anderson, and Genevieve Konopka

MET Receptor Tyrosine Kinase as an Autism Genetic Risk Factor
Yun Peng, Matthew Huentelman, Christopher Smith, and Shenfeng Qiu

Transcriptional Dysregulation of Neocortical Circuit Assembly in ASD
Kenneth Y. Kwan

Motor Skill in Autism Spectrum Disorders: A Subcortical View
Leanne Chukoskie, Jeanne Townsend, and Marissa Westerfield

Orchestration of Neurodevelopmental Programs by RBFOX1: Implications for Autism Spectrum Disorder
Brent R. Bill, Jennifer K. Lowe, Christina T. DyBuncio, and Brent L. Fogel

Immune Dysregulation in Autism Spectrum Disorder
Elaine Y. Hsiao

Autism Susceptibility Genes and the Transcriptional Landscape of the Human Brain
Shingo Miyauchi and Irina Voineagu

INDEX

Volume 114

Modern Concepts of Focal Epileptic Networks
Premysl Jiruska, Marco de Curtis, and John G.R. Jefferys

Neocortical Focus: Experimental View
Igor Timofeev, Sylvain Chauvette, and Sara Soltani

Malformations of Cortical Development and Neocortical Focus
Heiko J. Luhmann, Werner Kilb, and Hans Clusmann

Limbic Networks and Epileptiform Synchronization: The View from the Experimental Side
Charles Behr, Margherita D'Antuono, Shabnam Hamidi, Rochelle Herrington, Maxime Lévesque, Pariya Salami, Zahra Shiri, Rüdiger Köhling, and Massimo Avoli

Limbic Networks: Clinical Perspective
Aylin Y. Reid and Richard J. Staba

Modern Concepts of Seizure Modeling
Christophe Bernard, Sebastien Naze, Timothée Proix, and Viktor K. Jirsa

Mechanisms of Ictogenesis
Thomas Blauwblomme, Premysl Jiruska, and Gilles Huberfeld

Seizure Termination
Frédéric Zubler, Andreas Steimer, Heidemarie Gast, and Kaspar A. Schindler

Epileptic Focus and Alteration of Metabolism
Jakub Otáhal, Jaroslava Folbergrová, Richard Kovacs, Wolfram S. Kunz, and Nicola Maggio

Modern Techniques of Epileptic Focus Localization
Lukas Martinkovic, Hrvoje Hecimovic, Vlastimil Sulc, Radek Marecek, and Petr Marusic

From Treatment to Cure: Stopping Seizures, Preventing Seizures, and Reducing Brain Propensity to Seize
Ivan Pavlov and Stephanie Schorge

INDEX

Volume 115

Environmental Alterations of Epigenetics Prior to the Birth
Chiao-Ling Lo and Feng C. Zhou

Transgenerational Epigenetics and Brain Disorders
Nadia Rachdaoui and Dipak K. Sarkar

The Epigenetic Landscape of Alcoholism
Harish R. Krishnan, Amul J. Sakharkar, Tara L. Teppen, Tiffani D.M. Berkel, and Subhash C. Pandey

Epigenetic Regulatory Mechanisms in Stress-Induced Behavior
Sumana Chakravarty, Salil Saurav Pathak, Swati Maitra, Nitin Khandelwal, Karisetty Bhanu Chandra, and Arvind Kumar

Epigenetics of Schizophrenia: An Open and Shut Case
David P. Gavin and Christina Floreani

Epigenetic Mechanisms in Autism Spectrum Disorder
Adrian Zhubi, Edwin H. Cook, Alessandro Guidotti, and Dennis R. Grayson

MicroRNAs and Ethanol Toxicity
Rajesh C. Miranda

INDEX

Volume 116

Introduction to Sequencing the Brain Transcriptome
Robert Hitzemann, Priscila Darakjian, Nikki Walter, Ovidu Iancu, Robert Searles, and Shannon McWeeney

Analysis Considerations for Utilizing RNA-Seq to Characterize the Brain Transcriptome
Christina Zheng, Sunita Kawane, Daniel Bottomly, and Beth Wilmot

Data Integration and Reproducibility for High-Throughput Transcriptomics
Michael Mooney and Shannon McWeeney

Coexpression and Cosplicing Network Approaches for the Study of Mammalian Brain Transcriptomes
Ovidiu Dan Iancu, Alexander Colville, Priscila Darakjian, and Robert Hitzemann

Splicing in the Human Brain
Ammar Zaghlool, Adam Ameur, Lucia Cavalier, and Lars Feuk

Understanding Complex Transcriptome Dynamics in Schizophrenia and Other Neurological Diseases Using RNA Sequencing
Xi Wang and Murray J. Cairns

The Central Role of Noncoding RNA in the Brain
Boris Guennewig and Antony A. Cooper

Genetics of Gene Expression in CNS
Robert W. Williams and Ashutosh K Pandey

Transcriptomic Changes in Brain Development
Allissa A. Dillman and Mark R. Cookson

Gene Expression in the Addicted Brain
Zhifeng Zhou, Mary-Anne Enoch, and David Goldman

RNA-Seq Reveals Novel Transcriptional Reorganization in Human Alcoholic Brain
Sean P. Farris and R. Dayne Mayfield

INDEX

Volume 117

Learning-Induced Structural Plasticity in the Cerebellum
Hiroshi Nishiyama

Cerebellar Mechanisms of Learning and Plasticity Revealed by Delay Eyelid Conditioning
Michael D. Mauk, Wenke Li, Andrei Khilkevich, and Hunter Halverson

Cerebellar Long-Term Potentiation: Cellular Mechanisms and Role in Learning
Giorgio Grasselli and Christian Hansel

The Ontogeny of Associative Cerebellar Learning
John H. Freeman

INDEX

Volume 118

Neuroimmune Mechanisms of Alcohol and Drug Addiction
Changhai Cui, David Shurtleff, and R. Adron Harris

Neuroimmune Pathways in Alcohol Consumption: Evidence from Behavioral and Genetic Studies in Rodents and Humans
Gizelle Robinson, Dana Most, Laura B. Ferguson, Jody Mayfield, R. Adron Harris, and Yuri A. Blednov

Fetal Alcohol Spectrum Disorders and Neuroimmune Changes
Paul D. Drew and Cynthia J.M. Kane

Role of Microglia in Regulation of Ethanol Neurotoxic Action
Lucy Chastain and Dipak K. Sarkar

Functions of the Chemokine Receptor CXCR4 in the Central Nervous System and Its Regulation by μ-Opioid Receptors
Bradley Nash and Olimpia Meucci

Discovery of a Novel Site of Opioid Action at the Innate Immune Pattern-Recognition Receptor TLR4
Jonathan Henry W. Jacobsen, Linda R. Watkins, and Mark R. Hutchinson

Neuroimmune Basis of Methamphetamine Toxicity
Jennifer M. Loftis and Aaron Janowsky

Marijuana Use Brain Immune Mechanisms
Guy A. Cabral and Melissa Jamerson

Interactions of HIV and Drugs of Abuse: The Importance of Glia and Host Genetic Factors
Kurt F. Hauser and Pamela E. Knapp

Neuroimmune Basis of Alcoholic Brain Damage
Fulton T. Crews and Ryan P. Vetreno

Converging Actions of Alcohol on Liver and Brain Immune Signaling
Gyongyi Szabo and Dora Lippai

Opportunities for the Development of Neuroimmune Therapies in Addiction
Lara A. Ray, Daniel Roche, Keith Heinzerling, and Steve Shoptaw

Use of Addictive Substances and NeuroHIV
Sulie L. Chang, Kaitlyn P. Connaghan, Yufeng Wei, and Ming D. Li

INDEX

Volume 119

Adenosine Receptor Neurobiology: Overview
Jiang-Fan Chen, Chien-fei Lee, and Yijuang Chern

Adenosine Receptor PET Imaging in Human Brain
Masahiro Mishina and Kiich Ishiwata

An Overview of Adenosine A_{2A} Receptor Antagonists in Parkinson's Disease
Peter Jenner

Mode of Action of Adenosine A_{2A} Receptor Antagonists as Symptomatic Treatment for Parkinson's Disease
Akihisa Mori

Adenosine Receptors and Dyskinesia in Pathophysiology
Masahiko Tomiyama

Clinical/Pharmacological Aspect of Adenosine A_{2A} Receptor Antagonist for Dyskinesia
Tomoyuki Kanda and Shin-ichi Uchida

Interaction of Adenosine Receptors with Other Receptors from Therapeutic Perspective in Parkinson's Disease
Nicolas Morin and Thérèse Di Paolo

Effects of the Adenosine A_{2A} Receptor Antagonist on Cognitive Dysfunction in Parkinson's Disease
Shin-ichi Uchida, Takako Kadowaki-Horita, and Tomoyuki Kanda

Clinical Nonmotor Aspect of A_{2A} Antagonist in PD Treatment
Masahiro Nomoto, Masahiro Nagai, and Noriko Nishikawa

Adenosine Receptors and Huntington's Disease
Chien-fei Lee and Yijuang Chern

Adenosine Receptors and Epilepsy: Current Evidence and Future Potential
Susan A. Masino, Masahito Kawamura, Jr., and David N. Ruskin

Adenosine Receptor Control of Cognition in Normal and Disease
Jiang-Fan Chen

Adenosine Receptors in Cerebral Ischemia
Alessia Melani, Anna Maria Pugliese, and Felicita Pedata

Roles of Adenosine and its Receptors in Sleep–Wake Regulation
Zhi-Li Huang, Ze Zhang, and Wei-Min Qu

Involvement of Adenosine A_{2A} Receptors in Depression and Anxiety
Koji Yamada, Minoru Kobayashi, and Tomoyuki Kanda

The Adenosine Neuromodulation System in Schizophrenia
Daniel Rial, Diogo R. Lara, and Rodrigo A. Cunha

INDEX

Volume 120

The Story of "Speed" from "Cloud Nine" to Brain Gain
Andrew Lees, Katrin Sikk, and Pille Taba

Amphetamine-Type Stimulants: The Early History of Their Medical and Non-Medical Uses
Nicolas Rasmussen

Miracle or Menace?
Mike Jay

Psychostimulants: Basic and Clinical Pharmacology
Andrew C. McCreary, Christian P. Müller, and Małgorzata Filip

Epigenetic Mechanisms of Psychostimulant-Induced Addiction
Anti Kalda and Alexander Zharkovsky

Experimental Models on Effects of Psychostimulants
Sulev Kõks

Neurologic Complications of Psychomotor Stimulant Abuse
Juan Sanchez-Ramos

Neurobehavioral Sequelae of Psychostimulant Abuse
Atbin Djamshidian

Neuropsychiatric Adverse Effects of Amphetamine and Methamphetamine
Jaanus Harro

"Addicted to Euphoria": The History, Clinical Presentation, and Management of Party Drug Misuse
Jenny Bearn and Matthew O'Brien

"Natural Amphetamine" Khat: A Cultural Tradition or a Drug of Abuse?
Nilesh B. Patel

Methcathinone "Kitchen Chemistry" and Permanent Neurological Damage
Katrin Sikk and Pille Taba

"Legal Highs" – An Emerging Epidemic of Novel Psychoactive Substances
Jolanta B. Zawilska

Psychostimulants and Artistic, Musical, and Literary Creativity
Iain Smith

Opium as a Literary Stimulant: The Case of Samuel Taylor Coleridge
Neil Vickers

INDEX

FSC
www.fsc.org
MIX
Papier | Fördert gute Waldnutzung
FSC® C083411

Kolibri 360 GmbH
Ferdinand-Jühlke-Straße 7
99095 Erfurt, Deutschland
produktsicherheit@kolibri360.de